A REVIEW OF ALGEBRA

HEIDI A. HOWARD

Florida Community College-Jacksonville

Addison
Wesley

Boston San Francisco New York
London Toronto Sydney Tokyo Singapore Madrid
Mexico City Munich Paris Cape Town Hong Kong Montreal

Reproduced by Addison-Wesley from camera-ready copy supplied by the author.

Copyright © 2002 Pearson Education, Inc.

ISBN   0-201-77347-3

8 9 10 BRG 08 07 06

To my husband and daughter,
who passionately love numbers

Contents

Chapter 5 Roots and Radicals; Complex Numbers

Answers to Selected Exercises

Preface

Much of the material in this text is based on examples and exercises from *Beginning Algebra*, Eighth Edition and *Intermediate Algebra*, Eighth Edition by Margaret L. Lial and John E. Hornsby. Their successful student-friendly approach in writing mathematics textbooks is appreciated by a wide student and instructor audience.

This text is designed to review basic algebra skills. Students who plan to take College Algebra or Precalculus will find mastery of these topics to be an essential component to their success. Also, this book can be used as a tool for those students who wish to build algebra skills as a natural progression from arithmetic.

The major topics in this text include the *Real Number System*, *Linear Equations*, *Problems Solving*, *Polynomials*, *Exponents*, *Rational Expressions*, *Roots*, *Radicals*, and *Complex Numbers*. Within each section, self-check exercises are included after examples. Students should find these exercises an effective way to tackle the concepts in small pieces, before moving on to the section exercises. Students will find the answers to these self-check exercises just prior to the section exercises. Each chapter also includes a summary, quick review, review exercises, as well as a twenty-question chapter test. Answers to odd-numbered exercises from chapter sections and review exercises are included. Also answers to both even- and odd-numbered exercises for the chapter text are included. A complete detailed solution manual to accompany this text is available separately.

I wish to acknowledge the contributions made by Tim Mogill, who did the accuracy check on all of the manuscript materials; Bobbie Lewis, who prepared the original outline of the topics; Joe Vetere, who gave great technical support; as well as senior editors Bill Poole and Anne Kelly.

H. A. Howard

CHAPTER 1: THE REAL NUMBER SYSTEM

1.1 | Basic Terms

This chapter reviews some of the basic symbols and rules that are studied in algebra.

Sets: A **set** is a collection of objects called the **elements** or **members** of the set. In algebra, the elements in a set are usually numbers. Set braces, { }, are used to enclose the elements. For example, 2 is an element of the set {1, 2, 3}.

A set can be defined either by listing or by describing its elements. For example,

$$S = \{\text{Oregon, Ohio, Oklahoma}\}$$

defines the set S by *listing* its elements. The same set might be *described* by saying that set S is the set of all states in the United States whose names begin with the letter "O."

Set S above has a finite number of elements. Some sets contain an infinite number of elements, such as

$$N = \{1, 2, 3, 4, 5, 6, \ldots\},$$

where the three dots show that the list continues in the same pattern. Set N is called the set of **natural numbers**, or **counting numbers**. A set containing no elements, such as the set of natural numbers less than 1, is called the **empty set**, or **null set**, usually written $\varnothing$. The empty set may also be written as { }.

To write the fact that 2 is an element of the set {0, 1, 2, 3}, we use the symbol $\in$ (read "is an element of"):

$$2 \in \{0, 1, 2, 3\}.$$

Two sets are equal if they contain exactly the same elements. For example,

$$\{1, 2\} = \{2, 1\},$$

because the sets contain the same elements. (Order doesn't matter.) On the other hand, $\{1, 2\} \neq \{0, 1, 2\}$ ($\neq$ means "is not equal to") since one set contains the element 0 while the other does not.

In algebra, letters called **variables** are often used to represent numbers or to define sets of numbers. For example, the statement x is a natural number between 1 and 9 defines the set {2, 3, 4, 5, . . . , 8}.

The notation $\{x \mid x$ is a natural number between 1 and 9$\}$ is an example of **set-builder notation**. We read this notation this way: the set of all elements x such that x is a natural number between 1 and 9.

EXAMPLE 1
List the elements in the set.

(a) $\{x \mid x$ is a natural number less than 4$\} = \{1, 2, 3\}$

(b) $\{y \mid y$ is one of the first five even natural numbers$\} = \{2, 4, 6, 8, 10\}$

(c) $\{z \mid z$ is a natural number at least 7$\} = \{7, 8, 9, 10, \ldots\}$

EXAMPLE 2

Use set-builder notation to describe the set.

(a) {1, 3, 5, 7, 9} = {$y \mid y$ is one of the first five odd natural numbers}

(b) {5, 10, 15, . . .} = {$d \mid d$ is a multiple of 5 greater than 0}

Self-Check 1

1. List the elements in the set, {$w \mid w$ is one of the first 6 odd natural numbers}.

2. Use set-builder notation to describe the set, {3, 6, 9, 12, . . .}.

Determine if the following are true or false.

3. $26 \notin \{2, 4, 6, 8, \ldots\}$ **4.** $\{t, a, p\} \neq \{p, a, t\}$

The Common Sets of Numbers: The following sets of numbers are used throughout algebra.

Natural Numbers or   {1, 2, 3, 4, 5, 6, 7, 8, . . .}
   Counting Numbers

Whole Numbers   {0, 1, 2, 3, 4, 5, 6, . . .}

Integers   {. . . , −3, −2, −1, 0, 1, 2, 3, . . .}

Rational Numbers $\left\{ \dfrac{p}{q} \,\middle|\, p \text{ and } q \text{ are integers, with } q \neq 0 \right\}$

Irrational Numbers $\left\{ x \mid x \text{ is a real number that is not rational} \right\}$

Real Numbers $\left\{ x \mid x \text{ is represented by a point on a number line} \right\}$

Examples of irrational numbers include most square roots, such as $\sqrt{7}, \sqrt{11}, \sqrt{2}$, and $\sqrt{5}$. (Some square roots *are* rational: $\sqrt{16} = 4$, $\sqrt{100} = 10$, and so on.) Another irrational number is π, the ratio of the circumference of a circle to its diameter. All irrational numbers are real numbers.

The relationships among these various sets of numbers are shown in Figure 1; in particular, the figure shows that the set of real numbers includes both the rational and the irrational numbers.

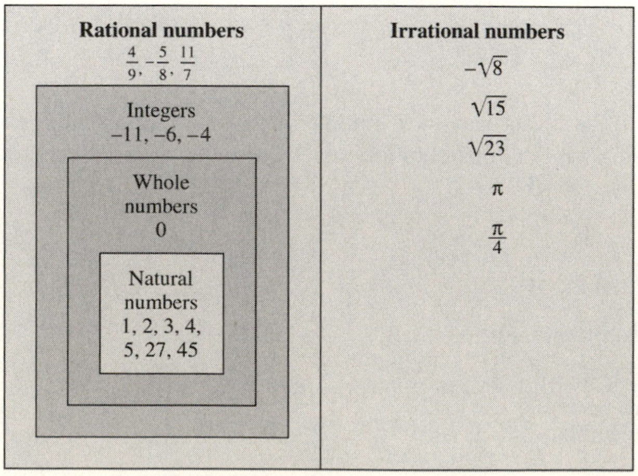

All numbers shown here are real numbers.

Figure 1

EXAMPLE 3

(a) $0, \frac{2}{3}, -\frac{9}{64}, \frac{28}{7}$ (or 4), and 2.45 are rational numbers.

(b) $\sqrt{3}, \pi, -\sqrt{2}$, and $\sqrt{7} + \sqrt{3}$, are irrational numbers.

(c) $-8, \frac{12}{2}, -\frac{3}{1}$, and $\frac{75}{5}$ are integers.

(d) All the numbers in parts (a), (b), and (c) above are real numbers.

(e) $\frac{4}{0}$ is undefined, since the definition of rational number requires the denominator to be nonzero. (However $\frac{0}{4}$ equals 0, which is a real number.)

Number Lines: Number lines provide a way to picture a set of numbers. To construct a **number line**, choose any point on a horizontal line and label it 0. Then choose a point to the right of 0 and label it 1. The distance from 0 to 1 establishes a scale that can be used to locate more points, with positive numbers to the right of 0 and negative numbers to the left of 0. A number line is shown in Figure 2.

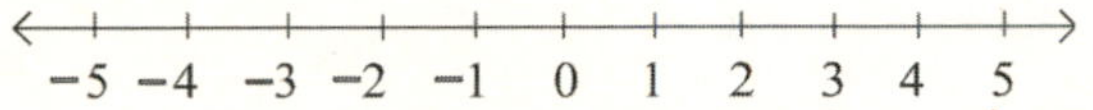

Figure 2

Each number is called the **coordinate** of the point that it labels, while the point is the **graph of the number**. A number line with several selected points graphed on it is shown in Figure 3.

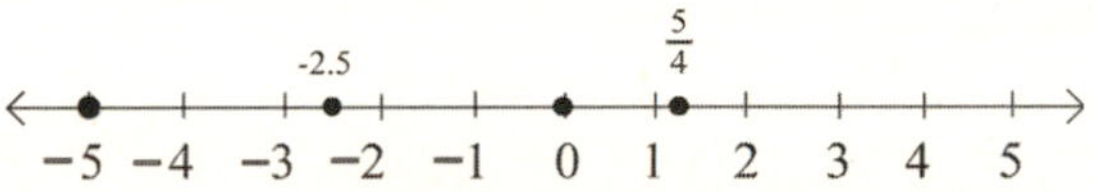

Figure 3

Additive Inverses: Two numbers that are the same distance from 0 on the number line but on opposite sides of 0 are called **additive inverses**, **negatives**, or **opposites** of each other. For example, 5 is the additive inverse of -5, and -5 is the additive inverse of 5.

Definition: Addition Inverse
For any number a, the number $-a$ is the additive inverse of a.

The number 0 is its own additive inverse. The sum of a number and its additive inverse is always 0 (the additive identity).

The symbol "$-$" can be used to indicate any of the following three things:
1. a negative number, such as -9 or -15;
2. the additive inverse of a number, as in "-4 is the additive inverse of 4";
3. subtraction, as in $12 - 3$.

In writing the number $-(-5)$, the symbol "$-$" is being used in two ways: the first $-$ indicates the additive inverse of -5, and the second indicates a negative number, -5. Since the additive inverse of -5 is 5, then $-(-5) = 5$. This example suggests the following property. For any number a,

$$-(-a) = a.$$

Absolute Value: The **absolute value** of a number a, written $|a|$, is the distance on the number line from 0 to a. For example, the absolute value of 5 is the same as the absolute value of -5, since each number lies five units from 0. See Figure 4. That is,

$$|5| = 5 \text{ and } |-5| = 5$$

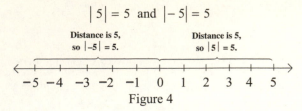

Figure 4

Since absolute value represents distance, and since distance is never negative, **the absolute value of a number is never negative**. The formal definition of absolute value is as follows.

Definition: Absolute Value

$$|a| = \begin{cases} a & \text{if } a \text{ is positive or } 0 \\ -a & \text{if } a \text{ is negative} \end{cases}$$

Notice that if a is negative, then $-a$ (the additive inverse of a) is a positive number and $|a|$ is positive.

EXAMPLE 4
Find the indicated value.

(a) $|2| = 2$ (b) $|-2| = 2$

(c) $|0| = 0$ (d) $-|8| = -(8) = -8$

(e) $-|-8| = -(8) = -8$ (f) $|-2| + |8| = 2 + 8 = 10$

EXAMPLE 5

     Absolute value is useful in applications comparing size without regard to sign. The U.S. House of Representative's 108th Congress was reapportioned according to the 2000 U.S. Census. For example, the number of seats in the House just prior to the Census and as a result of the Census for certain states is given in the following table.

State	Just prior to the 2000 U.S. Census	After the 2000 U.S. Census	Approximate Percent Rate of Change
Tennessee	10	9	−10.0
Pennsylvania	21	19	−9.5
Florida	23	25	+8.6
Texas	30	32	+6.6

Source: U.S. Census Bureau, January 2001

Which state had the greatest change in their representation? Which had the least change?

     We want the greatest change without regard to whether the change is an increase or a decrease. Look for the number in the Change column with the largest absolute value. That number is found in Tennessee, since $|-10.0| = 10.0$. Similarly, the least change is in Texas: $|6.6| = 6.6$.

Self-Check 2

Find the indicated value.

1. the additive inverse of −8.

2. $-|7|$

3. $|-13|$

4. $|12| - |-1|$

Inequality Symbols: A statement that two numbers are *not* equal is called an **inequality**. For example, the numbers $\frac{5}{7}$ and 3 are not equal. Write this inequality as

$$\frac{5}{7} \neq 3.$$

When two numbers are not equal, one must be less than the other. The symbol < means "is less than." For example, $-6 < 15$ and $0 < \frac{4}{3}$. Similarly, "is greater than" is written with the symbol >. For example, $9 > -2$ and $\frac{6}{5} > 0$. The number line in Figure 5 shows the graphs of the numbers 4 and 9. The graphs show that $4 < 9$.

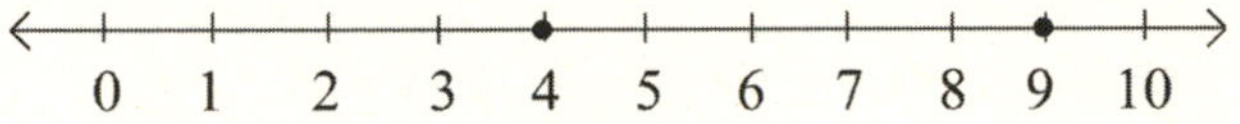

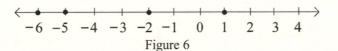

Figure 5

On the number line, the lesser of two given numbers is always located to the left of the other. $a < b$ if a is to the left of b. Also, if a is less than b, then b is greater than a. $b > a$ if b is to the right of a.

We can use a number line to determine order. As shown on the number line in Figure 6, -6 is located to the left of 1. For this reason, $-6 < 1$. Also, $1 > -6$. From the same number line, $-5 < -2$, or $-2 > -5$.

Figure 6

In addition to the symbols < and >, the symbols ≤ and ≥ often are used.

The symbol ≤ means "is less than or equal to." So $6 \leq 7$ and $7 \leq 7$. The symbol ≥ means "is greater than or equal to." So $-4 \geq -5$ and $-5 \geq -5$.

Self-Check 3

Determine if the statement is true or false.

1. $-4 < -8$

2. $\dfrac{1}{3} > 0.33$

Fill in the blank with the correct inequality symbol, either < or >.

3. $5 \underline{\quad} -2$

4. $-8 \underline{\quad} -18$

Self-Check Answers
1.1 {1, 3, 5, 7, 9, 11}
1.2 {$m \mid m$ is a natural number and a multiple of 3}
1.3 False   **1.4** False
2.1 8   **2.2** -7 **2.3** 13   **2.4** 11
3.1 False   **3.2** True   **3.3** >   **3.4** >

1.1 EXERCISES

For exercises 1-4, decide whether each statement about real numbers is true or false. If it is false, tell why.

1. Every number has a positive additive inverse.

2. Every number has a positive absolute value.

3. The absolute value of a negative number is its own additive inverse.

4. -11, $\dfrac{9}{7}$, and 0 are all rational numbers.

Write each set by listing its elements in exercises 5-10.

5. $\{y \mid y$ is a natural number greater than 4$\}$

6. $\{x \mid x$ is a natural number greater than or equal to 7$\}$

7. $\{z \mid z$ is an integer less than or equal to 3$\}$

8. $\{a \mid a$ is an even integer greater than 10$\}$

9. $\{k \mid k$ is an odd integer less than 3$\}$

10. $\{x \mid x$ is an irrational number that is also rational$\}$

11. A student claimed that $\{x \mid x$ is a natural number greater than 5$\}$ and $\{y \mid y$ is a natural number greater than 5$\}$ actually name the same set, even though different variables are used. Was the student correct?

12. A student claimed that $\{\varnothing\}$ and $\varnothing$ name the same set. Was this student correct?

Write each set using set-builder notation in exercises 13-16. (More than one description is possible.)

13. $\{3, 6, 9, 12, . . .\}$ 14. $\{. . ., -4, -2, 0, 2, 4, . . .\}$

15. $\{1, 3, 5, 7\}$ 16. $\{10, 11, 12, 13\}$

For exercises 17-18, which elements of the given set are (a) natural numbers, (b) whole numbers, (c) integers, (d) rational numbers, (e) irrational numbers, (f) real numbers, (g) undefined?

17. $\left\{-8,\ -\sqrt{5},\ -0.6,\ 0,\ \dfrac{1}{0},\ \dfrac{3}{4},\ \sqrt{5},\ 5,\ 7,\ \dfrac{17}{2},\ 17,\ \dfrac{60}{2}\right\}$

18. $\left\{-9,\ -\sqrt{6},\ -0.9,\ 0,\ \dfrac{2}{0},\ \dfrac{8}{7},\ \sqrt{3},\ 3,\ 8,\ \dfrac{31}{2},\ 19,\ \dfrac{75}{5}\right\}$

Graph the elements of the set on a number line in exercises 19-22.

19. $\{-3,-2,\ 0,\ 4,\ 8\}$

20. $\{-4,-3,\ 0,\ 3,\ 7\}$

21. $\left\{-\dfrac{2}{3},\ 0,\ \dfrac{7}{5},\ \dfrac{13}{5},\ \dfrac{11}{2},\ 4.6\right\}$

22. $\left\{-\dfrac{7}{5},\ -\dfrac{3}{4},\ 0,\ \dfrac{5}{6},\ \dfrac{13}{4},\ 5.5,\ \dfrac{9}{2}\right\}$

Find the value of each expression in exercises 23-34.

23. $|-7|$

24. $|-13|$

25. $-|7|$

26. $-|12|$

27. $-|-1|$

28. $-|-15|$

29. $-|4.9|$

30. $-|11.2|$

31. $|-1|+|6|$

32. $|-17|+|14|$

33. $|-8|-|-5|$

34. $|-12|-|-1|$

Refer to a number line to answer true or false for each statement in exercises 35-38.

35. $-8 < -2$

36. $-6 < -3$

37. $-0.32 < -\dfrac{5}{3}$

38. $-\dfrac{5}{3} < -1.1$

39. An inequality of the form "$a < b$" may also be written "$b > a$." Write $-4 < 2$ using this alternate form, and explain why both inequalities are true.

40. If $x > 0$ is a false statement for a given value of x, then is $x < 0$ necessarily a true statement? If not, explain why.

Use an inequality symbol to write each statement in exercises 41-48.

41. 6 is less than 12.

42. -5 is less than 13.

43. 5 is greater than x.

44. 7 is greater than z.

45. $5t-7$ is less than or equal to 10.

46. $6x+7$ is greater than or equal to 22.

47. 4 is greater than or equal to 4.

48. -2 is less than or equal to -2.

The wind chill index combines the temperature and wind speed to tell you how cold the wind makes it "feel." Wind chill, which is given as a temperature, is a guide to give you information as to how to dress properly when going outside. The table gives selected temperatures within states on a given day along with the temperatures with wind chill.

State	Temperature	Temperature with Wind Chill
Montana	-3^0	-19^0
Texas	41^0	39^0
Maine	21^0	21^0
Illinois	20^0	-10^0
Oklahoma	28^0	13^0
South Carolina	47^0	36^0

49. List the states in order, starting with the coldest and ending with the warmest according to the temperature (without wind chill).

50. List the states in order, starting with the coldest and ending with the warmest according to the temperature with wind chill.

51. Which state is affected the most by wind chill?

52. Which state is affected the least by wind chill?

1.2   Operations on Real Numbers

In this section we review the rules for adding, subtracting, multiplying, and dividing real numbers.

Add Real Numbers: The answer to an addition problem is called its **sum**. The rules for adding real numbers follow.

Rule for Adding Real Numbers

To add two numbers with the *same* sign, add their absolute values. The sign of the answer (either + or −) is the same as the sign of the two numbers.

To add two numbers with *different* signs, subtract the absolute values of the numbers. The answer is positive if the positive number has the larger absolute value. The answer is negative if the negative number has the larger absolute value.

EXAMPLE 1

(a) To add −11 and −4, first find their absolute values,

$$|-11| = 11 \quad \text{and} \quad |-4| = 4.$$

Since these numbers have the *same* sign, add their absolute values: $11 + 4 = 15$. Give the sum the sign of the two numbers. Since both numbers are negative, the sign is negative and

$$-11 + (-4) = -15.$$

(b) Find $-14 + 8$. Since these numbers have different signs, subtract the absolute values.

$$\left|-14\right| = 14 \quad \text{and} \quad \left|8\right| = 8$$
$$14 - 8 = 6$$

Give the result the sign of the number with the larger absolute value.

$$-14 + 8 = -6$$

Negative since $\left|-14\right| > \left|8\right|$

EXAMPLE 2
Add.

(a) $(-5) + (-3) = -(5 + 3) = -8$

(b) $(-17) + (-5) = -(17 + 5) = -22$

(c) $7 + (-4) = 3$

(d) $-12 + 17 = 5$

(e) $-\dfrac{1}{8} + \dfrac{5}{6} = -\dfrac{3}{24} + \dfrac{20}{24} = \dfrac{17}{24}$

(f) $-21 + 7 = -14$

Definition of Subtraction

For all real numbers a and b,

$$a - b = a + (-b).$$

(Change the sign of the second number and add.) The answer to a subtraction problem is called the **difference**.

EXAMPLE 3
Subtract.

Change to addition.

Change sign of second number.

(a) $7 - 12 = 7 + (-12) = -5$

Change to addition.

Sign changed.

(b) $-5 - 9 = -5 + (-9) = -14$

(c) $-9 - (-3) = -9 + \left[-(-3)\right]$
$$= -9 + 3$$
$$= -6$$

   For a problem with both addition and subtraction, add and subtract in order from the left, as in the following example. Remember to work inside parentheses or brackets first.

EXAMPLE 4

Perform the operations.

(a) $12 - (-5) - 7 - 4 = (12 + 5) - 7 - 4$
$$= 17 - 7 - 4$$
$$= 10 - 4$$
$$= 6$$

(b) $-11 - \left[-7 - (-6)\right] + 4 = -11 - \left[-7 + 6\right] + 4$
$$= -11 - \left[-1\right] + 4$$
$$= -11 + 1 + 4$$
$$= -10 + 4$$
$$= -6$$

Self-Check 1

Find the indicated value.

1. $(-14) + (-10)$ 2. $11 - 28$

3. $-5 - (-16)$ 4. $(-15) + (-13) - 25$

The Distance Between Two Points: If you consider two points on the number line, in order to determine the distance between them you could count up the number of units between them. By locating 3 and 6 on the number line in Figure 7 you can see that there are 3 units between. By locating −5 and 3 you can see that they have 8 units between them.

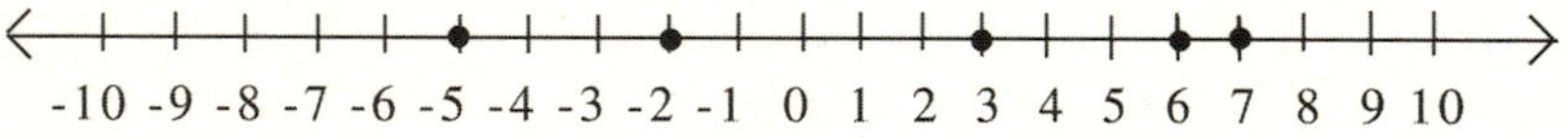

Figure 7

The **distance** between two points on a number line is the absolute value of the difference between the numbers.

EXAMPLE 5

Find the distance between the following pairs of points from Figure 7.

(a) The distance between 7 and −5 is 12 because

$$\left|7 - (-5)\right| = 12 \quad \text{or} \quad \left|-5 - 7\right| = 12.$$

(b) The distance between −5 and −2 is 3 because

$$\left|-5 - (-2)\right| = 3 \quad \text{or} \quad \left|-2 - (-5)\right| = 3.$$

Self-Check 2

Find the distance between the pairs of points.

1. 10 and 6   2. −7 and 5

3. −4 and 7   4. −6 and −14

Multiply Real Numbers: A **product** is the answer to a multiplication problem.   For example, 48 is the product of 6 and 8. The rules for products of real numbers follow.

Rule for Multiplying Real Numbers

The product of two numbers with the *same* sign is positive.

The product of two numbers with *different* signs is negative.

E X A M P L E 6
Multiply.

(a) $-4(-5) = 20$

(b) $-\dfrac{4}{5}\left(-\dfrac{7}{4}\right) = \dfrac{7}{5}$

(c) $9(-8) = -72$

▦(d) $-0.15(0.3) = -0.045$

Dividing Real Numbers: Multiplication and division are related operations. Division of two numbers can be restated as multiplication. For example,

$$\frac{36}{9} = 4 \text{ since } 36 = 9 \cdot 4.$$

The result of dividing two numbers is called the **quotient**.

Remember that division by 0 is always undefined. However, dividing 0 by a nonzero number gives a quotient of 0.

Here are some examples.

$$\frac{0}{9} = 0. \qquad \frac{11}{0} \text{ is undefined}. \qquad \frac{0}{0} \text{ is undefined}.$$

Self-Check 3

Multiply or divide as indicated.

1. $(34)(-17)$

2. $-\dfrac{2}{9} \div \left(-\dfrac{7}{12}\right)$

▦ **3.** $(0.2)(-4.1)$

▦ **4.** $(-0.40147) \div (0.8452)$

Exponents: A **factor** of a given number is any number that divides evenly (without remainder) into the given number. For example, 2 and 12 are factors of 24 since $2 \cdot 12 = 24$.

In algebra, exponents are used as a way of writing the products of repeated factors. For example, the product $3 \cdot 3 \cdot 3 \cdot 3 \cdot 3$ is written

$$3 \cdot 3 \cdot 3 \cdot 3 \cdot 3 = 3^5$$

The number 5 shows that 3 appears as a factor five times. The number 5 is the **exponent**, 3 is the **base**, and 3^5 is an **exponential expression**. Multiplying out the five 3s gives

$$3^5 = 3 \cdot 3 \cdot 3 \cdot 3 \cdot 3 = 243.$$

A number multiplied by itself is called the **square** of the number. 64 is "8 squared" because $8 \cdot 8 = 8^2 = 64$.

Reciprocals: If $b \neq 0$, then $\frac{1}{b}$ is the **reciprocal** of b. When a nonzero number and it's reciprocal are multiplied, the product is always 1 (the multiplicative identity). This fact is the reason behind the rule for division; that is, why we "multiply by the reciprocal of the denominator" when a division problem involves fractions.

$$\frac{a}{b} = \frac{a \cdot \frac{1}{b}}{b \cdot \frac{1}{b}} = \frac{a \cdot \frac{1}{b}}{1} = a \cdot \frac{1}{b}$$

Rule for Dividing Real Numbers

If a and b are real numbers and $b \neq 0$, then

$$\frac{a}{b} = a \cdot \frac{1}{b}.$$

EXAMPLE 7
Find the quotient.

(a) $\dfrac{12/5}{-6} = \dfrac{12}{5}\left(-\dfrac{1}{6}\right) = -\dfrac{2}{5}$

(b) $\dfrac{-\frac{4}{5}}{-\frac{1}{4}} = -\dfrac{4}{5}\left(-\dfrac{4}{1}\right) = \dfrac{16}{5}$

Notice that the rules for the sign of a quotient are the same as for the sign of a product.

The rules for multiplication and division suggest that certain forms of a fraction are equivalent. The fractions $\dfrac{-x}{y}$, $-\dfrac{x}{y}$, and $\dfrac{x}{-y}$ are equivalent. Also, the fractions $\dfrac{x}{y}$ and $\dfrac{-x}{-y}$ are equivalent. (Assume $y \neq 0$.)

EXAMPLE 8
Evaluate the exponential expression.

(a) $6^2 = 6 \cdot 6 = 36$

(b) $(-1)^2 = (-1) \cdot (-1) = 1$
    Base is -1.

(c) $2^6 = 2 \cdot 2 \cdot 2 \cdot 2 \cdot 2 \cdot 2 = 64$

(d) $-1^2 = -(1^2) = -(1) = -1$
    Base is 1.

Self-Check 4

Identify the base and exponent.

1. $(-4)^5$

2. -8^9

Write in exponential form.

3. $(-7)(-7)(-7)(-7)(-7)(-7)$

4. $\dfrac{4}{5} \cdot \dfrac{4}{5} \cdot \dfrac{4}{5}$

Square Roots: The opposite of squaring a number is called taking its **square root**. For example, a square root of 25 is 5. Another square root of 25 is −5, since $(-5)^2 = 25$. Thus, 25 has two square roots, 5 and −5. The positive square root of a number is written with the symbol $\sqrt{}$. For example, the positive square root of 25 is written $\sqrt{25} = 5$. The negative square root of 25 is written $-\sqrt{25}$. Since the square of any nonzero real number is positive, a number like is $\sqrt{-25}$ not a real number. The symbol $\sqrt{}$ is used only for the *positive* square root, except that $\sqrt{0} = 0$.

EXAMPLE 9
Find each root.

(a) $\sqrt{36} = 6$ since 6 is positive and $6^2 = 36$.

(b) $\sqrt{0} = 0$ since $0^2 = 0$.

(c) $\sqrt{\dfrac{25}{16}} = \dfrac{5}{4}$ since $\frac{5}{4}$ is positive and $\left(\frac{5}{4}\right)^2 = \frac{25}{16}$.

(d) $\sqrt{121} = 11$ since 11 is positive and $11^2 = 121$.

(e) $-\sqrt{121} = -11$

(f) $\sqrt{-121}$ is not a real number.

Self-Check 5

Simplify.

1. $\sqrt{64}$ 2. $\sqrt{\dfrac{25}{81}}$ 3. $-\sqrt{121}$ 4. $\sqrt{-144}$

The Order of Operations: Given a problem such as $4 + 7 \cdot 5$, should 4 and 7 be added first or should 7 and 5 be multiplied first? When a problem involves more than one operation, we use the following order of operations.

Procedure: Order of Operations

If no parentheses or brackets are present:

Step 1 Evaluate all powers and roots.

Step 2 Do any multiplications or divisions in the order in which they occur, working from left to right.

Step 3 Do any additions or subtractions in the order in which they occur, working from left to right.

EXAMPLE 10
Simplify.

(a) $4 + 7 \cdot 5$

To simplify this expression, first multiply and then add.

$$4 + 7 \cdot 5 = 4 + 35 \qquad \text{Multiply.}$$
$$= 39 \qquad \text{Add.}$$

(b) $24 \div 4 \cdot 5 + 7$

Multiplications and divisions are done in the order they appear from left to right. So divide first.

$$
\begin{aligned}
24 \div 4 \cdot 5 + 7 &= 6 \cdot 5 + 7 && \text{Divide.} \\
&= 30 + 7 && \text{Multiply.} \\
&= 37 && \text{Add.}
\end{aligned}
$$

If parentheses, square brackets, and fraction bars are present, work separately above and below any fraction bar. Work then within each set of parentheses or brackets. Start with the innermost set and work outward.

EXAMPLE 11

Simplify $2 \cdot 3^2 + 8 - (3 + 5)$.

Work inside the parentheses first.

$$2 \cdot 3^2 + 8 - (3 + 5) = 2 \cdot 3^2 + 8 - 8$$

Simplify powers and roots. Since $3^2 = 3 \cdot 3 = 9$,

$$2 \cdot 3^2 + 8 - 8 = 2 \cdot 9 + 8 - 8.$$

Do all multiplications or divisions, working from left to right.

$$2 \cdot 9 + 8 - 8 = 18 + 8 - 8$$

Finally, do all additions or subtractions, working from left to right.

$$
\begin{aligned}
18 + 8 - 8 &= 26 - 8 \\
&= 18
\end{aligned}
$$

EXAMPLE 12

Simplify $\frac{1}{3} \cdot 12 + (8 \div 2 \cdot 5)$.

Work inside the parentheses first, doing division before the multiplication.

$$
\begin{aligned}
\frac{1}{3} \cdot 12 + (8 \div 2 \cdot 5) &= \frac{1}{3} \cdot 12 + (4 \cdot 5) && \text{Divide in parentheses.} \\
&= \frac{1}{3} \cdot 12 + 20 && \text{Multiply in parentheses.} \\
&= 4 + 20 && \text{Multiply} \\
&= 24 && \text{Add}
\end{aligned}
$$

Self-Check 6

Simplify.

1. $18 - 6 \div 3$　　　　　　　　　　2. $(18 - 6) \div 3$

3. $5 \cdot (-3)^2 - 6 \cdot (-1) + 14$　　　4. $3 - (5 - 7)^3 \div (6 - 8) + 1$

Self-Check Answers

1.1 −24   **1.2** −17   **1.3** 11   **1.4** −53
2.1 4   **2.2** 12   **2.3** 11   **2.4** 8
3.1 −578   **3.2** $\frac{8}{21}$ **3.3** −0.82   **3.4** −0.475
4.1 base: -4; exponent: 5   **4.2** base: 8; exponent: 9
4.3 $(-7)^6$ **4.4** $\left(\frac{4}{5}\right)^3$
5.1 8   **5.2** $\frac{5}{9}$ **5.3** −11   **5.4** not a real number
6.1 16   **6.2** 4   **6.3** 65   **6.4** 0

1.2 EXERCISES

Complete each statement and give an example in exercises 1-10.

1. The sum of a positive number and a negative number is 0 if _____.

2. The sum two positive numbers is a _____ number.

3. The sum of two negative numbers is a _____ number.

4. The sum of a negative number and a positive number is a negative if _____.

5. The sum of a negative number and a positive number is a positive if ___.

6. The difference between two positive numbers is negative if _____.

7. The difference between two negative numbers is negative if _____.

8. The product of two numbers with like signs is _____.

9. The product of two numbers with unlike signs is _____.

10. The quotient formed by any nonzero number divided by 0 is _____, and the quotient formed by 0 divided by any nonzero number is _____.

Add or subtract as indicated in exercises 11-24.

11. $9 + (-5)$

12. $18 + (-12)$

13. $-7 + (-11)$

14. $-7 + (-17)$

15. $-\dfrac{4}{5} + \dfrac{5}{7}$

16. $-\dfrac{5}{6} + \dfrac{4}{9}$

17. $-0.137 + 0.212$

18. $-0.165 + 0.725$

19. $-9 - (-11) - (3 - 5)$

20. $-5 + (-12) + (-7 + 3)$

21. $\left(-\dfrac{7}{4} - \dfrac{1}{3}\right) + \dfrac{5}{6}$

22. $\left(-\dfrac{7}{8} + \dfrac{1}{4}\right) - \left(-\dfrac{1}{4}\right)$

23. $(-0.326) + (6 - 0.7)$

24. $(4 - 3.96) - (-0.75)$

The sketch shows a number line with several points labeled. Find the distance between each pair of points in exercises 25-28.

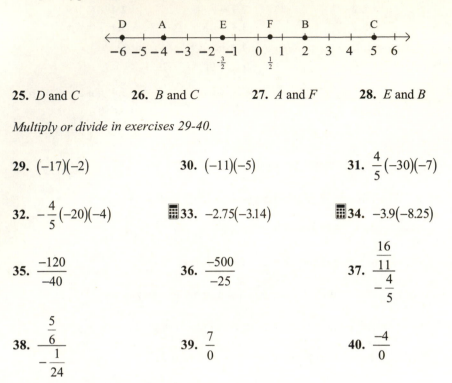

25. *D* and *C*   **26.** *B* and *C*   **27.** *A* and *F*   **28.** *E* and *B*

Multiply or divide in exercises 29-40.

29. $(-17)(-2)$ **30.** $(-11)(-5)$ **31.** $\frac{4}{5}(-30)(-7)$

32. $-\frac{4}{5}(-20)(-4)$ **33.** $-2.75(-3.14)$ **34.** $-3.9(-8.25)$

35. $\frac{-120}{-40}$ **36.** $\frac{-500}{-25}$ **37.** $\frac{\frac{16}{11}}{-\frac{4}{5}}$

38. $\frac{\frac{5}{6}}{-\frac{1}{24}}$ **39.** $\frac{7}{0}$ **40.** $\frac{-4}{0}$

For exercises 41-50, decide whether each statement is true or false. If the statement is false, explain why.

41. $(-3)^9$ is a negative number.   **42.** $(-3)^8$ is a positive number.

43. The product of 10 positive factors and 10 negative factors is positive.

44. The product of 5 positive factors and 5 negative factors is positive.

45. $-5^6 = (-5)^6$ **46.** $-5^7 = (-5)^7$

47. $\sqrt{25}$ is a positive number.   **48.** $1 + 2 \cdot 3 = 1 + (2 \cdot 3)$

49. In the exponential $-4^5, -4$ is the base.

50. $\sqrt{a}$ is positive for all positive numbers *a*.

Evaluate exercises 51-58.

51. $\sqrt{144}$ **52.** $\sqrt{361}$ **53.** 0.25^3 **54.** 0.85^3

55. $\left(-\frac{3}{10}\right)^2$ **56.** $-\left(\frac{3}{10}\right)^2$ **57.** $-\sqrt{100}$ **58.** $-\sqrt{400}$

59. Why is it incorrect to say that $\sqrt{36}$ is equal to 6 or −6?

60. Explain why $\sqrt{-1000}$ is not a real number.

For exercises 61-64, find the following roots on a calculator. Show as many digits as your calculator displays.

61. $\sqrt{12,144}$ **62.** $\sqrt{123.4}$ **63.** $\sqrt{76.18}$ **64.** $\sqrt{5.17}$

65. (a) If a is a positive number, is $-\sqrt{-a}$ positive, negative, or not a real number?

(b) If a is a positive number, is $-\sqrt{a}$ positive, negative, or not a real number?

66. Explain the rules for order of operations in your own words.

Perform each operation where possible, using the order of operations in exercises 67-78.

67. $-8(-2) - (-3)^3$

68. $-7 - 3(-5) + 4^2$

69. $\left| -8 - 7 \right| (-9) + 4^2$

70. $(-7 - 2) \left| -3 - 8 \right|$

71. $(-9 - 5)(-2 - 4)$

72. $\dfrac{(-11 + 5) \cdot (-6)}{-4 - 5}$

73. $\dfrac{(-9 + 3) \cdot (-8)}{-4 - 2}$

74. $\dfrac{4(-6 + 2)}{-2^2} - \dfrac{\left(-3^2 + 3\right)5}{2 - (-4)}$

75. $\dfrac{3(-8) + (-4)\left(-3^2\right)}{-8 + 3 + 5}$

76. $\dfrac{4(-9) + (-1)(-8)}{3^2 - 2 + (-7)}$

77. $\dfrac{6 - 4\left(\dfrac{-7 - 9}{-8}\right) - 2}{-10 - 13 + 4 \cdot 6}$

78. $\dfrac{-3\left(\dfrac{12 - (-20)}{4 \cdot 3 + 4}\right) - 4(-2 - 8)}{-7 - (-7) - \left[-6 - (-5)\right]}$

Use the table of Resident Population and Rank of selected States to answer questions in exercises 79 thru 82.

State	Population as of April 1, 1990	Population as of April 1, 2000	Rank as of April 1, 1990	Rank as of April 1, 2000
Alaska	550,043	626,932	49	48
Maine	1,227,928	1,274,923	38	40
Nevada	1,201,833	1,998,257	39	35
Ohio	10,847,115	11,353,140	7	7
West Virginia	1,793,477	1,808,344	34	37

Source: U.S. Census Bureau, January 2001.

79. Which State had the biggest change in population from April 1, 1990 to April 1, 2000? Was this change an increase in population or a decrease?

80. Which State had the biggest change in rank from April 1, 1990 to April 1, 2000? Was this change an increase in rank or a decrease?

81. Which State had the least change in population from April 1, 1990 to April 1, 2000?

82. Which State had the least change in rank from April 1, 1990 to April 1, 2000?

1.3 │ Properties of Real Numbers

In this section we discuss basic properties of real numbers. Understanding properties that govern the real numbers are helpful in performing calculations as well as simplifications of algebraic expressions.

The Distributive Property: Notice that

$$3(4 + 5) = 3 \cdot 9 = 27 \quad \text{and} \quad 3 \cdot 4 + 3 \cdot 5 = 12 + 15 = 27.$$

This example suggests the **distributive property of multiplication with respect** to addition, or simply the **distributive property**.

Distributive Property

For any real numbers *a, b,* and *c,*

$$a(b + c) = ab + ac \quad \text{and} \quad (b + c)a = ba + ca.$$

The distributive property can also be written

$$ab + ac = a(b + c).$$

This property is important because it provides a way to change a *product,* $a(b + c)$, to a *sum,* $ab + ac$, or a sum to a product. When the form $a(b + c) = ab + ac$ is used, we sometimes refer to it as "removing parentheses" or "expanding."

EXAMPLE 1
Use the distributive property to rewrite the expression.

(a) $5(x + y)$

   In the statement of the property, let $a = 5$, $b = x$, and $c = y$. Then

$$5(x + y) = 5x + 5y.$$

(b) $-3(7 + k) = -3(7) + (-3)(k) = -21 - 3k$

(c) $6x + 7x = (6 + 7)x = 13x$ (d) $2x - 8x = (2 - 8)x = -6x$

(e) $4x + 3y$

   Since there is no common factor here, we cannot use the distributive property to simplify the expression.

As illustrated in example 1(d), the distributive property can also be used for subtraction, so that

$$a(b - c) = ab - ac.$$

Self-Check 1

Use the distributive property to rewrite the expression.

1. $4(p + q)$ **2.** $7(x - y)$ **3.** $2r + 7r$ **4.** $12x - 17x$

The Inverse Properties: We saw that the additive inverse of a number a is $-a$. For example, 5 and -5 are additive inverses, as are -7 and 7. The number 0 is its own additive inverse. We also saw that two numbers with a product of 1 are reciprocals. Another name for a reciprocal of a number is its **multiplicative inverse**. Thus, 9 and $\frac{1}{9}$ are multiplicative inverses, and so are $-\frac{4}{5}$ and $-\frac{5}{4}$. (A pair of reciprocals has the same sign.) These properties are called the **inverse properties** of addition and multiplication.

Inverse Properties

For any real number a, there is a single real number $-a$, such that

$$a + (-a) = 0 \quad \text{and} \quad -a + a = 0.$$

For any nonzero real number a, there is a single real number $\frac{1}{a}$ such that

$$a \cdot \frac{1}{a} = 1 \quad \text{and} \quad \frac{1}{a} \cdot a = 1.$$

Self-Check 2

Find the additive inverse and multiplicative inverse for the following.

1. 2   **2.** -1 **3.** $\dfrac{3}{2}$ **4.** $-\dfrac{4}{9}$

The Identity Properties: The numbers 0 and 1 each have a special property. Zero is the only number that can be added to any number to get that number. That is, adding 0 leaves the identity of a number unchanged. For this reason, 0 is called the **identity element for addition**. In a similar way, multiplying by 1 leaves the identity of any number unchanged, so 1 is the **identity element for multiplication**. The following **identity properties** summarize this discussion.

Identity Properties

For any real number a,

$$a + 0 = 0 + a = a \quad \text{The identity property for 0}$$

and

$$a \cdot 1 = 1 \cdot a = a \quad \text{The identity property for 1}$$

The identity property for 1 is often used to simplify algebraic expressions.

EXAMPLE 2

Rewrite the expression using the identity property.

(a) $6m + m = 6m + 1m$ Identity property

$= (6 + 1)m$ Distributive property

$= 7m$

(b) $z + z = 1z + 1z$ Identity property

$= (1 + 1)z$ Distributive property

$= 2z$

(c) $-(p - 7q) = -1(p - 7q)$ Identity property

$= -1 \cdot p + (-1)(-7q)$ Distributive property

$= -p + 7q$

Self-Check 3

Rewrite the expression using the identity property and the distributive property.

1. $13y + y$ **2.** $x - 14x$ **3.** $-r - r$ **4.** $-(3p + q)$

Expressions such as $6m$ and $7q$ from Example 2 are examples of *terms*. A **term** is a number or the product (or quotient) of a number and one or more variables raised to powers. Terms with exactly the same variables raised to exactly the same powers are called **like terms**. The number in the product is called the **numerical coefficient** or just the **coefficient**. For example, in the term $9p$, the coefficient is 9.

The Commutative and Associative Properties: In the expression,

$$8m + 9 - 3m,$$

there are three terms. Notice that $(8m)(9)(-3m)$ is just one term, since it is a *product* of numbers and variables.

Simplifying expressions as in Examples 2(a) and (b) is called **combining like terms**. Only like terms may be combined. To combine like terms in the expression

$$8m + 9 - 3m$$

we need two more properties.

Commutative Properties

For any real numbers a and b,

$$a + b = b + a \text{ and } ab = ba.$$

For example,

$$5 + 2 = 2 + 5 \text{ and } 3(-4) = (-4)(3).$$

Associative Properties

For any real numbers *a, b,* and *c,*

$$a + (b + c) = (a + b) + c \text{ and } a(bc) = (ab)c.$$

For example,

$$4 + (3 + 2) = (4 + 3) + 2 \quad \text{and} \quad -6(2 \cdot 7) = (-6 \cdot 2) \cdot 7.$$

The associative properties are used to *regroup* the terms of an expression.   The commutative properties are used to change the *order* of the terms in an expression.

EXAMPLE 3

Combine like terms in $8m + 9 - 3m$.

By the order of operations, the first step would be to add $8m$ and 9, but they are unlike terms. To get $8m$ and $-3m$ together, use the associative and commutative properties. Begin by inserting parentheses and brackets according to the order of operations.

$$8m + 9 - 3m = (8m + 9) + (-3m)$$

$$= 8m + (9 + (-3m)) \qquad \text{Associative property}$$

$$= 8m + ((-3m) + 9) \qquad \text{Commutative property}$$

$$= (8m + (-3m)) + 9 \qquad \text{Associative property}$$

$$= (8 + (-3))m + 9 \qquad \text{Distributive property}$$

$$= 5m + 9$$

Together the properties of real numbers allow us to simplify expressions.

EXAMPLE 4
Simplify.

(a) $6x^2 - 4x^2 - 17x^2 + 22x^2 = (6 - 4 - 17 + 22)x^2 = 7x^2$

(b) $-3(y - 4) = -3(y) - (-3)(4) = -3y - (-12) = -3y + 12$

(c) $4x^2 + 7 - 6(x^2 + 3) - 9$

   First use the distributive property to eliminate the parentheses.

$$4x^2 + 7 - 6(x^2 + 3) - 9 = 4x^2 + 7 - 6x^2 - 18 - 9$$

Next use the commutative and associative properties to rearrange the terms; then combine like terms.

$$= 4x^2 - 6x^2 + 7 - 18 - 9$$

$$= -2x^2 - 20$$

(d) $9 - (4r + 6)$

   Think of $9 - (4r + 6)$ as $9 - 1(4r + 6)$.

$$9 - 1(4r + 6) = 9 - 4r - 6$$

$$= 3 - 4r$$

Self-Check 4

Combine like terms.

1. $6y + 2 - 8y - 18$

2. $7x^2 + 3 - 16x^2 - 8$

Simplify the expression.

3. $[(2x - 3) - 7x] + 6$

4. $3p - 4(6p - 8)$

The Multiplication Property of 0: The additive identity property gives a special property of 0, namely that $a + 0 = a$ for any real number a. The **multiplication property of 0** gives a special property of 0 that involves multiplication: The product of any real number and 0 is 0.

Multiplication Property of 0

For any real numbers a,

$$a \cdot 0 = 0 \quad \text{and} \quad 0 \cdot a = 0.$$

Self-Check Answers

1.1 $4p + 4q$ **1.2** $7x - 7y$ **1.3** $9r$ **1.4** $-5x$

2.1 additive inverse: -2; multiplicative inverse: $\frac{1}{2}$

2.2 additive inverse: 1; multiplicative inverse: -1

2.3 additive inverse: $-\frac{3}{2}$; multiplicative inverse: $\frac{2}{3}$

2.4 additive inverse: $\frac{4}{9}$; multiplicative inverse: $-\frac{9}{4}$

3.1 $14y$ **3.2** $-13x$ **3.3** $-2r$ **3.4** $-3p - q$

4.1 $-2y - 16$ **4.2** $-9x^2 - 5$ **4.3** $-5x + 3$ **4.4** $-21p + 32$

1.3 EXERCISES

Choose the correct response in exercises 1-3.

1. The identity element for addition is

    (a) $-a$.   (b) 0.   (c) 1.   (d) $\dfrac{1}{a}$.

2. The identity element for multiplication is

    (a) $-a$.   (b) 0.   (c) 1.   (d) $\dfrac{1}{a}$.

3. The coefficient in the term $-4xy^3$ is

    (a) -4.   (b) x.   (c) y^2.   (d) $-4x$.

Complete each statement in exercises 4-6.

4. Like terms are defined to be _____.

5. The distinction between the commutative and associative properties is that _____.

6. The multiplication property of 0 states that _____.

Use the properties of real numbers to simplify each expression in exercises 7-20.

7. $4x + 7x$ **8.** $9a + 4a$ **9.** $-10r + 4r$

10. $-7n + 10n$ **11.** $-8p + 12q$ **12.** $-16x + 4y$

13. $-p + 9p$ **14.** $-s + 12s$ **15.** $3(p + q)$

16. $5(a + b)$ **17.** $-13(x - y)$ **18.** $-8(p - q)$

19. $-6(3d + f)$ **20.** $-3(4m + n)$

Simplify each expression by removing parentheses and combining like terms in exercises 21-32.

21. $5x + 8x + 10 + 9$ **22.** $7m + 12m + 7 + 11$

23. $-14y + 5y + 8 + 7y$ **24.** $-6r - 10r + 5r - 2$

25. $4(k + 5) - 7k + 2 + 10$ **26.** $3(r - 5) + 7r - 6r + 5$

27. $0.25(12 + 16p) - 0.5(8 + 2p)$ **28.** $0.4(20 - 5x) - 0.6(5 + 10x)$

29. $-(3p + 9) + 4(3p + 2) - p$ **30.** $-(6m - 11) - 3(3m + 5) - 9m$

31. $1 + 4(3z - 7) - 2(4z + 6) - 15$ **32.** $12 + 8(z - 5) - 2(5z + 6) - 8$

For exercises 33-46, complete each statement so that the indicated property is illustrated. Simplify the answer, if possible.

33. $2x + 10x = $_____ (distributive property)   **34.** $8y - 2y = $_____ (distributive property)

35. $7(8r) = $_____ (associative property)   **36.** $-5 + (11 + 9) = $_____ (associative property)

37. $7x + 14y = $_____ (commutative property)   **38.** $-2 \cdot 9 = $_____ (commutative property)

39. $1 \cdot 8 = $_____ (identity property)   **40.** $-11x + 0 = $_____ (identity property)

41. $-\frac{3}{4}tz + \frac{3}{4}tz = $_____ (inverse property)   **42.** $-\frac{12}{7}\left(-\frac{7}{12}\right) = $_____ (inverse property)

43. $9(-3 + x) = $_____ (distributive property)   **44.** $2(x + y - z) = $_____ (distributive property)

45. $0(0.8296x + 17y + 15z) = $_____ (multiplication property of 0)

46. $0\left(32t^3 - 10t + 15\right) =$ _____

$\qquad\qquad\qquad$ (multiplication property of 0)

47. Give an "everyday" example of the commutative operative and of one that is not commutative.

48. Give an "everyday" example of inverse operations.

49. Replace x with 6 and show that $5 + 6x \neq 11x$.

50. Replace x with 6 and show that $5x - x \neq 5$.

Use the distributive property to calculate the following values mentally in exercises 51-56.

51. $81 \cdot 15 + 19 \cdot 15$ $\qquad\qquad$ **52.** $18 \cdot 60 + 18 \cdot 40$

53. $64 \cdot \dfrac{3}{2} - 14 \cdot \dfrac{3}{2}$ $\qquad\qquad$ **54.** $9.75(14) - 9.75(4)$

55. $5.21(68) + 5.21(32)$ $\qquad\qquad$ **56.** $\dfrac{5}{4}(27) + \dfrac{5}{4}(13)$

While it may seem that simplifying the expression $9x + 4 + 2x + 5$ to $11x + 9$ is fairly easy, there are several important steps that require mathematical justification. These steps are usually done mentally. For now, provide the property that justifies the statement in the simplification in exercises 57-62. (These steps could be done in other orders.)

57. $8x + 5 + 2x + 12 = (8x + 5) + (2x + 12)$

58. $\qquad\qquad\qquad = 8x + (5 + 2x) + 12$

59. $\qquad\qquad\qquad = 8x + (2x + 5) + 12$

60. $\qquad\qquad\qquad = (8x + 2x) + (5 + 12)$

61. $\qquad\qquad\qquad = (8 + 2)x + (5 + 12)$

62. $\qquad\qquad\qquad = 10x + 17$

63. The statement $7 - 9 = 9 - 7$ is a false example. In general, subtraction is not a commutative operation. Under what conditions would $a - b = b - a$?

64. By the distributive property, $a(b + c) = ab + ac$. This property is more completely named the distributive property of multiplication with respect to addition. Is there a distributive property of multiplication with respect to multiplication? That is, does

$$a(bc) = (ab)(ac)?$$

for all real numbers a, b, and c? To find out, try various sample values of a, b, and c.

1.4 | Variables, Expressions, and Equations

Evaluating Expressions: A **variable** is a symbol, usually a letter, such as x, y, or z, used to represent any unknown number. An **algebraic expression** is a collection of numbers, variables, operation symbols, and grouping symbols (such as parentheses). For example,

$$7(m + 3), \quad 8x - 9, \quad \text{and} \quad 9c^2 + 7c + 1$$

are all algebraic expressions. In the algebraic expression $8x - 9$, the term $8x$ indicates the product of 8 and x, just as $9c^2$ shows the product of 9 and c^2. Also, $7(m + 3)$ means the product of 7 and $m + 3$. An algebraic expression has different numerical values for different values of the variable.

EXAMPLE 1
Find the value of each expression when $x = 4$ and $y = 7$.

(a) $3x + 2y$

Replace x with 4 and y with 7. Follow the order of operations; multiply first, then add.

$$
\begin{aligned}
3x + 2y &= 3 \cdot 4 + 2 \cdot 7 && \text{Let } x = 4 \text{ and } y = 7. \\
&= 12 + 14 && \text{Multiply.} \\
&= 26 && \text{Add.}
\end{aligned}
$$

(b) $\dfrac{3x - 6y}{5x - 2y}$

Replace x with 4 and y with 7.

$$
\begin{aligned}
\frac{3x - 6y}{5x - 2y} &= \frac{3 \cdot 4 - 6 \cdot 7}{5 \cdot 4 - 2 \cdot 7} && \text{Let } x = 4 \text{ and } y = 7. \\
&= \frac{12 - 42}{20 - 14} && \text{Multiply.} \\
&= \frac{-30}{6} && \text{Subtract.} \\
&= -5 && \text{Divide.}
\end{aligned}
$$

Self-Check 1

Find the value of the expression when $x = 2$ and $y = 3$.

1. $2x - 5y$ **2.** $\dfrac{3x - 2y}{x + y}$

Find the value of the expression when $p = -4$ and $q = 8$.

3. $2p + 7q$ **4.** $\dfrac{4p}{q - p}$

Distinguish Between an Expression and an Equation: Students often have trouble distinguishing between equations and expressions. Remember that an equation is a sentence; an expression is a phrase.

$$
\begin{array}{cc}
2x + 8 = 10 & \qquad 2x + 8 \\
\uparrow & \qquad \uparrow \\
\text{Equation} & \qquad \text{Expression} \\
\text{(to solve)} & \qquad \text{(to simplify or evaluate)}
\end{array}
$$

EXAMPLE 2
Decide whether each of the following is an equation or an expression.

(a) $3x - 5y$ There is no equal sign, so this is an expression.

(b) $3x = 5y$ Because of the equal sign, this is an equation.

Translate Words and Phrases Involving Addition and Subtraction: Word phrases can be translated into algebraic expressions. These translations are used in problem solving. We begin by working with addition words and phrases.

The word *sum* is one of the words that indicated addition. The chart below lists some of the words and phrases that also signify addition.

Word or Phrase	Example	Numerical Expression
Sum of	The **sum of** −2 and 9	$-2 + 9$
Added to	7 **added to** −12	$-12 + 7$
More than	15 **more than** −18	$-18 + 15$
Increased by	−9 **increased by** 7	$-9 + 7$
Plus	16 **plus** 6	$16 + 6$

EXAMPLE 3
Write a numerical expression for each phrase, and simplify the expression.

(a) The **sum of** −8 and 14 and 12

$$-8 + 14 + 12 = 6 + 12 = 18$$

   Add in order from left to right.

(b) 4 **more than** −2, **increased by** 5

$$(-2 + 4) + 5 = 2 + 5 = 7$$

Self-Check 2

Write a numerical expression for each phrase, and simplify the expression.

1. 16 more than −7

2. The sum of 14 and −16, increased by 12

3. −14 added to 9

4. 5 plus 8, increased by 1

We now look at how we interpret words and phrases that involve subtraction. In order to solve problems that involve subtraction, we must be able to interpret key words and phrases that indicate subtraction. *Difference* is one of them. Some of these are given in the chart below.

Word or Phrase	Example	Numerical Expression
Difference between	The **difference between** −7 and 5	$-7 - 5$
Subtracted from	5 **subtracted from** 13	$13 - 5$
Less	7 **less** 1	$7 - 1$
Less than	15 **less than** 7	$7 - 15$
Decreased by	8 **decreased by** −5	$8 - (-5)$
Minus	10 **minus** 6	$10 - 6$

When you are subtracting two numbers, it is important that you write then in the correct order, because, in general $a - b \neq b - a$. For example, $7 - 3 \neq 3 - 7$. For this reason, think carefully before interpreting an expression that involves subtraction. (This difficulty did not rise for addition because addition has the commutative property.)

EXAMPLE 4
Simplify the expression.

(a) The **difference between** −2 and 9
It is conventional to write the numbers in the order they are given when "difference between" is used.

$$-2 - 9 = -2 + (-9) = -11$$

(b) 6 **subtracted from** the sum of 7 and −8
Here addition is also used, as indicated by the word *sum*. First add 7 and −8. Next, subtract 6 *from* this sum.

$$\left[7 + (-8)\right] - 6 = -1 - 6 = -1 + (-6) = -7$$

(c) 7 **less than** −12
Be careful with order here. 7 must be taken *from* −12.

$$-12 - 7 = -12 + (-7) = -19$$

Notice that "7 less than −12" differs from "7 *is less than* −12." The second of these is symbolized as $7 < -12$ (which is a false statement).

(d) 3, **decreased by** 8 **less than** 11
First, write "8 less than 11" as $11 - 8$. Next, subtract $11 - 8$ from 3.

$$3 - (11 - 8) = 3 - 3 = 0$$

Self-Check 3

Write a numerical expression for each phrase, and simplify the expression.

1. 12 subtracted from −2

2. The difference between 1 and −16, increased by 4

3. −18 less 3   **4.** 7 less 8, increased by 1

Translate Words and Phrases Involving Multiplication and Division: Just as there are words and phrases that indicate addition and subtraction, certain ones also indicate multiplication and division.

The word *product* refers to multiplication. The chart below gives other key words or phrases that indicate multiplication.

Word or Phrase	Example	Numerical Expression
Product of	The **product of** −9 and −2	$(-9)(-2)$
Times	17 **times** −3	$17(-3)$
Twice (meaning "2 times")	**Twice** 5	$2(5)$
Of (used with fractions)	$\frac{1}{2}$ **of** 30	$\frac{1}{2}(30)$
Percent of	10% **of** 80	$0.10(80)$

EXAMPLE 5

Write a numerical expression for each phrase and simplify. Use the order of operations.

(a) The **product of** 13 and the sum of 5 and −8

Here 13 is multiplied by "the sum of 5 and −8."

$$13\left[5 + (-8)\right] = 13(-3) = -39$$

(b) **Twice** the difference between 7 and −12

$$2\left[7 - (-12)\right] = 2\left[7 + 12\right] = 2(19) = 38$$

(c) Four-thirds of the sum of −8 and −7

$$\frac{4}{3}\left[-8 + (-7)\right] = \frac{4}{3}\left[-15\right] = -20$$

Self-Check 4

Write a numerical expression for each phrase, and simplify the expression.

1. The product of 6 and −2   **2.** Twice the sum of 6 and −18

3. 20% of −18 less 3   **4.** 10 less than 8, times 4

The word *quotient* refers to the answer in a division problem. In algebra, quotients are usually represented with a fraction bar; the symbol ÷ is seldom used. When translating applied problems involving division, use a fraction bar. The chart gives some key phrases associated with division.

Phrase	Example	Numerical Expression
Quotient of	The *quotient of* −12 and 3	$\frac{-12}{3}$
Divided by	−24 *divided by* −8	$\frac{-24}{-8}$
Ratio of	The *ratio of* 7 to 5	$\frac{7}{5}$

It is customary to write the first number names as the numerator and the second as the denominator when interpreting a phrase involving division, as shown in the next example.

EXAMPLE 6

Write a numerical expression for each phrase, and simplify the expression.

(a) The **quotient of** 18 and the sum of −11 and 2

"Quotient" indicated division. The number 18 is the numerator and "the sum of −11 and 2" is the denominator.

$$\frac{18}{-11 + 2} = \frac{18}{-9} = -2$$

(b) The product of 8 and −7, **divided by** the difference of 6 and −2

The numerator of the fraction representing the division is obtained by multiplying 8 and −7. The denominator is found by subtracting 6 and −2.

$$\frac{8(-7)}{6 - (-2)} = \frac{-56}{6 + 2} = \frac{-56}{8} = -7$$

Self-Check 5

Write a numerical expression for each phrase, and simplify the expression.

1. The quotient of 6 and -2 **2.** The sum of 2 and -16, divided by 7

3. The ratio of -18 and 4   **4.** The product of 5 and 4 divided by -2

Variables and Expressions: In algebra, we often use variables to translate key words and phrases into mathematical expressions. Here is a summary of common verbal expressions and their translation. x represents the unknown number.

Verbal Expression	Mathematical Expression
Addition	
The **sum of** a number and 8	$x + 8$
9 **more than** a number	$x + 9$
3 **plus** a number	$3 + x$
67 **added to** a number	$x + 67$
A number **increased by** 6	$x + 6$
The **sum of** two numbers	$x + y$
Subtraction	
4 **less than** a number	$x - 4$
16 **minus** a number	$16 - x$
A number **decreased by** 7	$x - 7$
The **difference between** two numbers	$x - y$
A number **subtracted from** 9	$9 - x$
Multiplication	
18 **times** a number	$18x$
Some number **multiplied by** 8	$x \cdot 8$ or more commonly written as $8x$
$\frac{4}{7}$ **of** some number (used only with fractions and percent)	$\frac{4}{7}x$
Twice (2 times) some number	$2x$
The **product of** two numbers	xy
Division	
The **quotient of** 7 and some number	$\frac{7}{x}\ (x \neq 0)$
A number **divided by** 17	$\frac{x}{17}$
The **ratio of** two numbers or the quotient of two numbers	$\frac{x}{y}\ (y \neq 0)$

Because subtraction and division are not commutative operations, it is important to correctly translate expressions involving them. For example, "4 less than a number" is translated as $x - 4$, *not* $4 - x$. "A number subtracted from 7" is expressed as $7 - x$, not $x - 7$. For division, it is understood that the number doing the dividing is the denominator, and the number that is divided is the numerator. For example, "a number divided by 11" and "11 divided into x" both translate as $\frac{x}{11}$. Similarly, "the quotient of x and y" is translated as $\frac{x}{y}$.

Self-Check 6

Write a mathematical expression for each phrase, and simplify them if possible.

1. The sum of a number and -3 **2.** Twice the sum of a number and 5

3. The quotient of 7 and some number, divided by 6

4. 15 less some number

Translate Sentences into Equations: The preceding table listed mathematical *expressions* that correspond to phrases. *Equations* correspond to sentences. To write an equation, look for two expressions that are equal. The word "is" often signals the translation to the mathematical symbol = ("equals"). In fact, any words that indicated the idea of sameness also can translate to =.

EXAMPLE 7
Translate the following verbal sentences into equations.

Verbal Sentences	Equation
Twice a number, decreased by 7, **is** 65.	$2x - 7 = 65$
The product of a number and 13 increased by 12 **results in** 142.	$13x + 12 = 142$
A number divided by the sum of 6 and the number **is** 12.	$\dfrac{x}{6 + x} = 12$
The quotient of a number and 6, plus the number **is** 14.	$\dfrac{x}{6} + x = 14$

EXAMPLE 8
Write the following in symbols, using x as the variable, and guess or use trial and error to find the solution. All solutions come from the list of integers between -12 and 12, inclusive.

(a) Four **times** a number **is** -24.
   The word *times* indicates multiplication, and the word *is* translates as the equals sign.

$$4x = -24$$

Since the integer between -12 and 12, inclusive, that makes this statement true is -6, the solution of the equation is -6.

(b) The **sum** of a number and 3 **is** 10.

$$x + 3 = 10$$

Since $7 + 3 = 10$, the solution of this equation is 7.

(c) The **difference between** a number and 7 **is** 0.

$$x - 7 = 0$$

Since $7 - 7 = 0$, the solution of this equation is 7.

(d) The **quotient of** 30 and a number **is** -3.

$$\frac{30}{x} = -3$$

Here, x must be a negative number, since the numerator is positive and the quotient is negative. Since $\frac{30}{-10} = -3$, the solution is -10.

Self-Check 7

Translate the following into symbols, using x as the variable, and guess or use trial and error to find the solution. All solutions come from the list of integers between -12 and 12, inclusive.

1. 5 more than a number is 10.

2. 42 divided by a number results in -6.

3. 6 times a number less that number is 25.

4. A number increased by three produces twice that number.

Self-Check Answers

1.1 -11 **1.2** 0 **1.3** 48 **1.4** $-\frac{4}{3}$

2.1 $-7 + 16 = 9$ **2.2** $14 + (-16) + 12 = -2 + 12 = 10$

2.3 $9 + (-14) = -5$ **2.4** $5 + 8 + 1 = 13 + 1 = 14$

3.1 $-2 - 12 = -2 + (-12) = -14$

3.2 $\left[1 - (-16)\right] + 4 = \left[1 + 16\right] + 4 = 17 + 4 = 21$

3.3 $-18 - 3 = -18 + (-3) = -21$ **3.4** $\left[7 - 8\right] + 1 = -1 + 1 = 0$

4.1 $6(-2) = -12$ **4.2** $2\left[6 + (-18)\right] = 2(-12) = -24$

4.3 $0.20(-18 - 3) = 0.20(-21) = -4.2$ **4.4** $(8 - 10) \cdot 4 = (-2)(4) = -8$

5.1 $\dfrac{6}{-2} = -3$ **5.2** $\dfrac{2 + (-16)}{7} = \dfrac{-14}{7} = -2$

5.3 $\dfrac{-18}{4} = -\dfrac{9}{2}$ **5.4** $\dfrac{5 \cdot 4}{-2} = \dfrac{20}{-2} = -10$

6.1 $x + (-3)$ **6.2** $2\left[x + 5\right] = 2x + 2(5) = 2x + 10$

6.3 $\dfrac{7 + x}{6}$ **6.4** $15 - x$

7.1 $x + 5 = 10; 5$ **7.2** $\dfrac{42}{x} = -6; -7$ **7.3** $6x - x = 25; 5$ **7.4** $x + 3 = 2x; 3$

1.4 EXERCISES

For exercises 1-16, find the numerical value if (a) $x = 1$ and $y = 2$ and (b) $x = 3$ and $y = -2$.

1. $9x + y + 2$ **2.** $5x + 7y + 9$ **3.** $2(x + 5y)$

4. $5(3x + y)$ **5.** $x + \dfrac{6}{y}$ **6.** $y + \dfrac{12}{x}$

7. $\dfrac{x}{3} + \dfrac{y}{4}$ **8.** $\dfrac{x}{5} + \dfrac{y}{6}$ **9.** $\dfrac{x + 2y - 8}{3y + 2}$

10. $\dfrac{5x + 3y - 2}{x}$ **11.** $3y^2 + 4x$ **12.** $5x^2 + 5y$

13. $\dfrac{x + 2y^2}{3x + 4y}$ **14.** $\dfrac{x^2 + 4}{-2x + 6y}$ **15.** $0.211x^2 + 0.58y^2$

16. $0.911x^2 + 0.411y^2$

For exercises 17-30, change each word phase to an algebraic expression. Use x as the variable to represent the number.

17. Five times a number

18. Eight times a number

19. Six added to a number

20. Twelve added to a number

21. Three subtracted from a number

22. Nine subtracted from a number

23. A number subtracted from six

24. A number subtracted from fifteen

25. The difference between a number and 5

26. The difference between 5 and a number

27. 14 divided by a number

28. A number divided by 14

29. The product of 8 and five less than a number

30. The product of 9 and six more than a number

31. In the phrase "Five more than the product of a number and 7," does the word *and* signify the operation of addition? Explain.

32. Suppose that the directions on a test read, "Solve the following expressions." How would you politely correct the person who wrote these directions? What alternative directions might you suggest?

For exercises 33-40, change each word statement to an equation. Use x as the variable. Find all solutions from the set {2, 4, 6, 8, 10}.

33. The sum of a number and 6 is 16.

34. A number minus five equals 1.

35. Twenty minus three-fourths of a number is 14.

36. The sum of seven-fifths of a number and 1 is 15.

37. One more than twice a number is 7.

38. The product of a number and 4 is 16.

39. Three times a number is equal to 10 more than twice a number.

40. Eighteen divided by a number equals $\frac{1}{2}$ times that number.

Identify as an expression *or an* equation *in exercises 41-46.*

41. $4x + 3(x - 9)$

42. $10y - (5y + 9)$

43. $14t + 3(t + 8) = 1$

44. $-8r + 2(r - 5) = 14$

45. $x + y = 5$

46. $x + y - 5$

Write a numerical expression for each phrase and simplify in exercises 47-54.

47. The sum of −7 and 11 and 8

48. The sum of −3 and 7 and −17

49. 12 added to the sum of −21 and−5

50. −5 added to the sum of −21 and 13

51. The sum of −8 and −12, increased by 12

52. The sum of −9 and −11, increased by 12

53. 6 more than the sum of 9 and −23

54. 10 more than the sum of −4 and −6

Write a numerical expression for each phrase and simplify in exercises 55-62.

55. The difference between 5 and −12

56. The difference between 9 and −11

57. 9 less than −4

58. 2 less than −62

59. The sum of 12 and −6, decreased by 9

60. The sum of 23 and −18, decreased by 12

61. 18 less than the difference between 9 and −3

62. 21 less than the difference between 11 and −9

Write a numerical expression for each phrase and simplify in exercises 63-76.

63. The product of −8 and 3, added to 7

64. The product of 5 and −9, added to −18

65. Twice the product of −1 and 8, subtracted from −5

66. Twice the product of −9 and 3, subtracted from −14

67. Eight subtracted from the product of 1.6 and −3.7

68. Nine subtracted from the product of 3.6 and −2.5

69. The product of 11 and the difference between 6 and −9

70. The product of −8 and the difference between 4 and −12 .

71. The quotient of −14 and the sum of −5 and 2

72. The quotient of −12 and the sum of −9 and −3

73. The sum of 16 and −4 , divided by the product if 6 and −2

74. The sum of −19 and −7 , divided by the product if 2 and −4

75. The product of $-\dfrac{1}{2}$ and $\dfrac{3}{8}$, divided by $-\dfrac{4}{3}$

76. The product of $-\dfrac{8}{3}$ and $-\dfrac{1}{4}$, divided by $\dfrac{1}{9}$

In exercises 77-84 write each statement in symbols, using x as the variable, and find the solution by guessing or by using trial and error. All solutions come from the set of integers between −12 and 12.

77. Seven times a number is −42.   **78.** Nine times a number is −27.

79. The quotient of a number and 2 is −5.   **80.** The quotient of a number and 5 is −1.

81. 3 less than a number is 5.   **82.** 8 less than a number is 3.

83. When 6 is added to a number, the result is −6.

84. When 3 is added to a number, the result is −3.

CH 1 | Summary

KEY TERMS

1.1 set — elements (members)
empty set (null set) — variable
set-builder notation — number line
coordinate — graph
additive inverse (negative, opposite) — signed numbers
absolute value — inequality
interval — interval notation

1.2 sum — difference — distance
product — quotient — factor
factored form — exponent — base
reciprocal — square of a number — square root

1.3 distributive property — multiplicative inverse
inverse properties — identity element for addition
identity element for multiplication — term
like terms — coefficient (numerical coefficient)
combining like terms — multiplication property of 0

1.4 algebraic expression — equation

CH 1 | Quick Review

1.1 BASIC TERMS

Sets of Numbers

Natural Numbers $\{1, 2, 3, 4, 5, 6, 7, 8, \ldots\}$

Whole Numbers $\{0, 1, 2, 3, 4, 5, 6, \ldots\}$

Integers $\{\ldots, -3, -2, -1, 0, 1, 2, 3, \ldots\}$

Rational Numbers $\left\{ \dfrac{p}{q} \,\middle|\, p \text{ and } q \text{ are integers, with } q \neq 0 \right\}$, or
$\{x \mid x \text{ has a terminating or repeating decimal representation}\}$

Irrational Numbers $\{x \mid x \text{ is a real number that is not rational}\}$, or
$\{x \mid x \text{ has a nonterminating, nonrepeating decimal representation}\}$

Real Numbers $\left\{ x \mid x \text{ is represented by a point on a number line} \right\}$

1.2 OPERATIONS ON REAL NUMBERS

Addition
Same signs: Add the absolute values. The sum has the same sign as the numbers.
Different signs: Subtract the absolute values. The answer has the sign of the number with the larger absolute value.

Subtraction
Change the sign of the second number and add.

Multiplication
Same signs: The product is positive.
Different signs: The product is negative.

The product of an even number of negative factors is positive.
The product of an odd number of negative factors is negative.

Division
Same signs: The quotient is positive.
Different signs: The quotient is negative.

Order of Operations
If parentheses, square brackets, or fraction bars are present:
Step 1 Work separately above and below any fraction bar.
Step 2 Use the rules that follow within each set of parentheses or square brackets. Start with the innermost set and work outward.

If no parentheses or brackets are present:
Step 1 Evaluate all powers and roots.
Step 2 Do any multiplications or divisions in the order in which they occur, working from left to right.
Step 3 Do any additions or subtractions in the order in which they occur, working from left to right.

1.3 PROPERTIES OF REAL NUMBERS
For any real numbers, a, b, and c:

Distributive Property
$a(b + c) = ab + ac$

Inverse Property
$a + (-a) = 0$ and $-a + a = 0$

$a \cdot \dfrac{1}{a} = 1$ and $\dfrac{1}{a} \cdot a = 1$ $(a \neq 0)$

Commutative Properties
$a + b = b + a$

$ab = ba$

Associative Properties
$a + (b + c) = (a + b) + c$

$a(bc) = a(bc)$

Multiplication Property of 0
$a \cdot 0 = 0$ and $0 \cdot a = 0$

1.4 VARIABLES, EXPRESSIONS, AND EQUATIONS
An expression is a phrase and an equation is a sentence. Key words and phrases can be translated into expressions and equations.

CH 1   Review Exercises

Graph each set on the number line in exercises 1-2.

1. $\left\{-5, -3, 3, \dfrac{5}{4}, 5\right\}$ **2.** $\left\{-4, -\dfrac{9}{4}, -0.75, 0, 4, \dfrac{11}{3}\right\}$

Find the value of each expression in exercises 3-5.

3. $|-17|$ **4.** $|13|$ **5.** $-|-8|$

Let set $S = \left\{-10, -\dfrac{5}{3}, -\sqrt{11}, 0, \dfrac{7}{3}, \sqrt{5}, \dfrac{8}{4}\right\}$. *Simplify the elements of S as necessary and then list the elements that belong to the specified set in exercises 6-9.*

6.   Whole numbers   **7.**   Integers

8.   Rational numbers   **9.**   Real numbers

Write each set by listing its elements in exercises 10-11.

10. $\{x \mid x \text{ is a natural number between 4 and 8}\}$

11. $\{y \mid y \text{ is a whole number less than 5}\}$

Write true *or* false *for each inequality in exercises 12-13.*

12. $3 \cdot 2 \leq |11 - 5|$ **13.** $3 + |-2| > 5$

The table below gives the unemployment rate in New York City on January for the years 1991-2000. A negative change means the unemployment rate has gone down from one year to the next (in January). A positive change means the rate has gone up.

Year	Rate of Unemployment in January	Change in Unemployment Rate
1991	7.8	
1992	10.6	+35.9%
1993	11.9	+12.3%
1994	10.4	−12.6%
1995	8.1	−22.1%
1996	8.8	+ 8.6%
1997	10.1	+14.8%
1998	9.1	− 9.9%
1999	7.8	−14.3%
2000	6.6	−15.4%

Source: Bureau of Labor Statistics, January 2001.

14. (a) In which years in New York City did the January rate of unemployment show the greatest change?
     (b) In which years in New York City did the January rate of unemployment show the least change?

Add or subtract as indicated in exercises 15-23.

15. $-\dfrac{2}{7} - \left(-\dfrac{3}{5}\right)$

16. $-\dfrac{7}{6} - \left(-\dfrac{1}{12}\right)$

17. $-6 + (-8) + 17 - 6$

18. $-9.25 + 6.21 - 8.5 - 2.7$

19. $-20 + (-12) + (-8)$

20. $-2 - 7 - (-12) + (-9)$

21. $-21 + (-8) + (-19)$

22. $-2 - 8 - (-11) + (-9)$

23. $\dfrac{2}{5} - \left(\dfrac{1}{10} - \dfrac{3}{2}\right)$

24. State in your own words how to determine the sign if the sum of two numbers.

25. How is subtraction related to addition?

Find each product or quotient in exercises 26-29.

26. $-2(-4)(-2)(-3)$ **27.** $-\dfrac{5}{7}\left(-\dfrac{21}{20}\right)$

28. $\dfrac{-51}{-17}$ **29.** $\dfrac{-6.39}{-0.71}$

Use the table given in exercise 14 to answer true *or* false *for each statement in exercises 30-32. Use absolute value for comparison.*

30. The absolute value of percent change from January 1998 to 1999 was greater than the absolute value in the change from January 1999 to 2000.

31. The largest percent change was from January 1991 to 1992.

32. The smallest percent change was from January 1999 to 2000.

33. Which one of the following is undefined? $\dfrac{5 - (-5)}{2 - 2}$ or $\dfrac{5 - 5}{2 - (-2)}$

Evaluated each expression in exercises 34-39. If it is not a real number, say so.

34. $\left(\dfrac{2}{3}\right)^3$ **35.** $(2.5)^2$

36. $\sqrt{900}$ **37.** $\sqrt{-100}$

38. $-20\left(\dfrac{4}{5}\right) + 12 \div 4$ **39.** $\dfrac{-4(2)^2 + 5\sqrt{16} - 6}{1 - 3\left(-\sqrt{9}\right)}$

Let $k = -2$, $m = 3$, and $n = 25$, and evaluate each expression in exercises 40-41.

40. $5k - 6m$ **41.** $-5\sqrt{n} + m - 2k$

42. In order to evaluate $(5 + 4)^2$, should you work within the parentheses first, or should you square 5 and square 4 and add?

43. By replacing a with 3 and b with 4, show that $(a + b)^2 \neq a^2 + b^2$.

Use properties of real numbers to simplify each expression in exercises 44-53.

44. $3x + 5x$ **45.** $19v - 14v$

46. $-k + 5k$ **47.** $8p - p$

48. $-3(k + 4)$ **49.** $7(r + 5)$

50. $-6y + 2 - 8 + 7y$ **51.** $3a + 6 - a - 2 - a - 5$

52. $-3(k - 1) + 5k - k$ **53.** $-2(2m - 1) + 3(3m - 1) - 4(m + 1)$

Complete each statement so that the indicated property is illustrated in exercises 54-62. Simplify the answer, if possible.

54. $3x + 4x = $ _____   **55.** $-5 \cdot 1 = $ _____
    (distributive property)   (identity property)

56. $3(2x) = $ _____   **57.** $-5 + 15 = $ _____
    (associative property)   (commutative property)

58. $-4 + 4 = $ _____   **59.** $3(x + y) = $ _____
    (inverse property)   (distributive property)

60. $0 + 9 = $ _____   **61.** $2 \cdot \dfrac{1}{2} = $ _____
    (identity property)   (inverse property)

62. $\dfrac{2}{17} \cdot 0 = $ _____
    (multiplication property of 0)

Perform the indicated operations in exercises 63-84.

63. $\left(-\dfrac{2}{5}\right)^2$

64. $-\dfrac{9}{25}(-50)$

65. $\dfrac{-60}{20}$

66. $8(3x + 2y)$

67. $10\left(\dfrac{2}{5}\right) + 2^2 - 8 \div \sqrt{4}$

68. $(-4)^3$

69. $-(2x - 5y)$

70. $-2 + |-8| + |-14|$

71. $-\sqrt{16}$

72. $\dfrac{2\sqrt{9} - 5\sqrt{16}}{-3 \cdot 2 + 8(-4) - 11}$

73. $-2.1(3.48)$

74. $-\dfrac{15}{11} \div -\dfrac{22}{25}$

75. $-\dfrac{3}{7}\left[6(-8) + 8 - \sqrt{81}\right]$

76. $-(a + 2b) - (3a - 7b)$

77. $-3(2x^2 + 3y)$ if $x = -2$ and $y = 7$

78. $2.3 - 5.7 - 2.1 + 3.14$

79. $-|-5| + |-21|$

80. -4^2

81. $-\dfrac{2.72}{0.34}$

82. $-\dfrac{3}{4} - \left(\dfrac{5}{8} - \dfrac{7}{12}\right)$

83. $-3x + 5 - 6x + 1$

84. $\dfrac{6m^2 - 5n^3}{3k^2 - 1}$ if $m = 3$, $n = 12$, and $k = -8$

CH 1 | Test

1. Graph $\left\{-4,\ -0.5,\ 2,\ \dfrac{7}{4},\ 5.25\right\}$ on the number line.

Let $A = \left\{-\sqrt{3},\ -5,\ -0.75,\ 0,\ 4,\ 6.9,\ \dfrac{15}{3}\right\}$. *First simplify each element as needed and then list the elements from A that belong to the set in exercises 2-3.*

2. Integers

3. Rational numbers

Perform the indicated operations in exercises 4-7.

4. $-8 + 12 + (-5) - (-9)$

5. $8 - 5^2 + 3(-4) + (-6)^2$

6. $\dfrac{2 - 3^2 + (-6)(2)}{\sqrt{16}(-3)}$

7. $\dfrac{5}{6} - \left(-\dfrac{1}{8} - \dfrac{7}{12}\right)$

Find each indicated root in exercises 8-9. If the number is not real, say so.

8. $-\sqrt{400}$

9. $\sqrt{-400}$

Let $k = -3$, $m = 5$, and $n = 4$, and evaluate each expression in exercises 10-11.

10. $-4\sqrt{n} + 5m - 7k$

11. $\dfrac{m^2 + 2n^2}{k^3 + 16}$

Use properties of real numbers to simplify each expression in exercises 12-15.

12. $x + 9x$

13. $2z - (-13z)$

14. $-2(y - 1) + 6y - y$

15. $-3(n - 4) + 2(3n - 2)$

16. How does the subtraction sign affect the terms $-9x$ and 7 when simplifying $(4x + 9) - (-9x + 7)$? What is the simplified form?

Match each statement with the appropriate property in exercises 17-20.

17. $3x + 2x = (3 + 2)x$

A. commutative property

18. $0 \cdot (-121) = 0$

B. associative property

19. $(2x + 3) + 4x = 2x + (3 + 4x)$

C. distributive property

20. $(2x + 3) + 4x = (3 + 2x) + 4x$

D. multiplication property of zero

CHAPTER 2: LINEAR EQUATIONS AND PROBLEM SOLVING

2.1 | The Addition and Multiplication Properties of Equality

In Chapter 1 we investigated rules of algebra and applied these rules to translate and rewrite expressions. These skills will all be applied in this chapter in the solving equations applying mathematics to real-world situations.

Linear Equations: The simplest type of equation is a *linear equation*.

Definition of a Linear Equation

A **linear equation** in one variable can be written in the form

$$Ax + B = 0$$

for real numbers A and B, with $A \neq 0$.

For example,

$$3x - 6 = 0, \quad 6x + 12 = -5, \quad \text{and} \quad x = 8$$

are linear equations; the last two can be written in the specified form using the properties to be developed in this section. However,

$$x^2 - 3x = 18, \quad \frac{1}{x} = 6x, \quad \text{and} \quad |5x + 2| = 4$$

are *not* linear equations.

An equation is solved by finding its **solution set**, the set of all solutions. Equations that have exactly the same solution sets are **equivalent equations**. To solve a linear equation, use a series of steps producing equivalent equations that yield an equation of the form $x = $ a number.

The Addition Property of Equality: The addition property of equality says that the same expression may be *added* to both sides of an equation.

Addition Property of Equality

If *A, B,* and *C* are mathematical expressions that represent real numbers, then the equations

$$A = B \quad \text{and} \quad A + C = B + C$$

have exactly the same solution.

EXAMPLE 1

Solve $x - 14 = 5$.

If the left side of this equation were just x, the solution would be found. Get x alone by using the addition property of equality and adding 14 on both sides.

$$x - 14 = 5$$
$$x - 14 + 14 = 5 + 14 \quad \text{Add 14 on both sides.}$$
$$x = 19$$

Check by substituting 19 for x in the original equation.

$$x - 14 = 5 \quad \text{Original Equation}$$
$$19 - 14 = 5 \ ? \quad \text{Let } x = 19.$$
$$5 = 5 \quad \text{True}$$

Since the check results in a true statement, {19} is the solution set.

EXAMPLE 2

Solve the equation $5k - 14 + k + 3 = 2 + 5k + 7$.

Begin by combining like terms on each side of the equation to get

$$6k - 11 = 9 + 5k.$$

Next, get all terms that contain variables on the same side of the equation and all terms without variables on the other side. One way to start is to subtract $5k$ from both sides.

$$6k - 11 - 5k = 9 + 5k - 5k \qquad \text{Subtract } 5k \text{ from both sides.}$$
$$k - 11 = 9 \qquad \text{Combine terms.}$$
$$k - 11 + 11 = 9 + 11 \qquad \text{Add 11 to both sides.}$$
$$k = 20$$

Check by substituting 20 for k in the original equation. The check results in a true statement, so the solution set is {20}.

Subtracting $5k$ in Example 2 is the same as adding $-5k$, since by the definition of subtraction $a - b = a + (-b)$. Therefore, the addition property also permits *subtracting* the same expression from both sides of an equation.

Self-Check 1

Solve the following equations.

1. $a - 6 = -2$ 2. $p + 2 = -11$

3. $7x + 3 - x - 10 = 3x + 2x - 4$ 4. $5z + 3z - 1 = 7z - 1$

The Multiplication Property of Equality: To solve an equation like $4x + 1 = 17$,

$$4x + 1 = 17$$
$$4x + 1 - 1 = 17 - 1 \qquad \text{Subtract 1 from both sides.}$$
$$4x = 16$$

another property is needed to change $4x = 16$ to $x =$ a number. The **multiplication property of equality** states that both sides of an equation can be multiplied by the same nonzero expression.

Multiplication Property of Equality

If A, B, and C are mathematical expressions that represent real numbers, then the equations

$$A = B \quad \text{and} \quad AC = BC$$

have exactly the same solution. (Assume that $C \neq 0$.)

Continuing from the equation above, use the multiplication property of equality to multiply both sides of the equation by $\frac{1}{4}$. Use $\frac{1}{4}$ because $4 \cdot \frac{1}{4} = \frac{4}{4} = 1$. The left-hand side will then be $1x$ or x, as follows:

$$4x = 16$$

$$\frac{1}{4}(4x) = \frac{1}{4} \cdot 16 \qquad \text{Multiply both sides by } \frac{1}{4}.$$

$$\left(\frac{1}{4} \cdot 4\right)x = \frac{1}{4} \cdot 16 \qquad \text{Associative property}$$

$$1x = 4 \qquad\qquad \text{Multiplicative inverse property}$$

$$x = 4 \qquad\qquad \text{Identity Property}$$

The solution set of the equation is {4}. Check by substituting 4 for x in the original equation.

Just as the addition property of equality permits subtracting the same number from both sides of an equation, the multiplication property of equality permits dividing both sides of an equation by the same nonzero number. For example, the equation $4x = 16$ could also be solved by dividing both sides by 4:

$$4x = 16$$

$$\frac{4x}{4} = \frac{16}{4} \qquad \text{Divide by 4.}$$

$$x = 4$$

EXAMPLE 3

Solve the $3.4x = 9.18$.

　　Divide both sides of the equation by 3.4.

$$\frac{3.4x}{3.4} = \frac{9.18}{3.4}$$

$$x = 2.7 \qquad \text{Divide.}$$

EXAMPLE 4

Solve $\frac{2}{5}h = 4$.

　　To get h alone, multiply both sides of the equation by $\frac{5}{2}$.

$$\frac{2}{5}h = 4$$

$$\frac{5}{2}\left(\frac{2}{5}h\right) = \frac{5}{2} \cdot 4 \qquad \text{Multiply by } \tfrac{5}{2}.$$

$$1 \cdot h = \frac{5}{2} \cdot \frac{4}{1}$$

$$h = 10$$

We could also have multiplied both sides of the equation by 5 to eliminate the denominator.

$$\frac{2}{5}h = 4$$

$$5\left(\frac{2}{5}h\right) = 5 \cdot 4 \qquad \text{Multiply by 5.}$$

$$2h = 20$$

$$\frac{2h}{2} = \frac{20}{2} \qquad \text{Divide by 2.}$$

$$h = 10$$

Using either method, the solution set is {10}. Check the answer by substitution in the original equation.

Use Properties to Solve a Linear Equation: The next example uses the distributive property to first simplify an equation.

EXAMPLE 5

Solve $2(3+7x)-(1+15x)=2$.

$$2(3 + 7x) - 1(1 + 15x) = 2 \qquad \text{Replace the } - \text{ sign with } -1.$$
$$6 + 14x - 1 - 15x = 2 \qquad \text{Distributive property}$$
$$5 - x = 2 \qquad \text{Combine like terms.}$$
$$5 - x - 5 = 2 - 5 \qquad \text{Subtract 5.}$$
$$-x = -3$$

The variable is alone on the left, but its coefficient is -1, since $-x = -1 \cdot x$. When this occurs, simply multiply both sides by -1.

$$-x = -3$$
$$-1 \cdot x = -3 \qquad\qquad -x = -1 \cdot x$$
$$-1(-1 \cdot x) = -1(-3) \qquad \text{Multiply by } -1.$$
$$x = 3$$

The solution set is {3}. Check by substituting into the original equation. From the final steps in Example 5, we can see that the following is true.

$$\text{If } -x = a, \quad \text{then } x = -a.$$

Self-Check 2

Solve the following equations.

1. $3.9x = 27.3$

2. $\dfrac{7}{4}p = 21$

3. $5(x+3)-2x-10=20$

4. $7(z+2)-9z-1=-5$

Self-Check Answers

1.1 4	**1.2** −13	**1.3** 3	**1.4** 0
2.1 7	**2.2** 12	**2.3** 5	**2.4** 9

2.1 EXERCISES

1. Which of the pairs of equations are equivalent equations?

    (a) $x+3=7$ and $x=4$ (b) $8-x=4$ and $x=-4$

    (c) $x+4=10$ and $x=6$ (d) $5+x=10$ and $x=-5$

2. In your own words, state the addition and multiplication properties of equality and give examples.

3. State how you would find the solution of a linear equation if your next-to-last step reads "$-x=3$."

For exercises 4-19, solve each equation using the addition property of equality. Check each solution.

4. $x+4=11$

5. $x-5=17$

6. $x-10.4=-6.7$

7. $y+13.1=-6.2$

8. $\frac{3}{5}w-8=\frac{8}{5}w$

9. $-\frac{1}{7}z+1=\frac{6}{7}z$

10. $5.2x+12=4.2x$

11. $9.8x-3=8.8x$

12. $8p+9=12+7p$

13. $11b-4=-9+10b$

14. $4.9y-9=3.9y-9$

15. $8.2r+15=7.2r+15$

16. $\frac{1}{8}x+6=-\frac{7}{8}x$

17. $\frac{1}{4}x-1=-\frac{3}{4}x$

18. $15x+6-14x=0$

19. $20x+2-19x=0$

20. Which of the following are not linear equations in one variable?
(a) $x^2-4x+3=0$
(b) $x^4=x$
(c) $2x-7=0$
(d) $5x-4x=7+8x$

Define a linear equation in one variable in words.

21. Refer to the definition of linear equation in one variable given in this section. Why is the restriction $A\neq 0$ necessary?

Solve each equation in exercises 22-39. First simplify both sides of the equation as much as possible. Check each solution.

22. $7t+5+2t-8t=2+11$

23. $2x+7x-7-8x=16+1$

24. $8x+4+5x+1=12x+2$

25. $5x-4-9x+3=-5x+12$

26. $9.8q-2.1-11.4q=-0.6q-2.1$

27. $-3.0x+3.5-1.4x=-5.4x+3.5$

28. $\frac{7}{9}x+\frac{5}{4}=\frac{2}{3}-\frac{2}{9}x+\frac{2}{3}$

29. $\frac{5}{6}s-\frac{2}{5}=\frac{4}{5}-\frac{1}{6}s+\frac{1}{6}$

30. $(7y+10)-(8+6y)=18$

31. $(5r-8)-(4r+2)=-9$

32. $2(p+4)-(6+p)=-4$

33. $5(k-4)-(4k+3)=-7$

34. $-5(2b+1)+(11b-8)=0$

35. $-6(3w-2)+(4+19w)=0$

36. $9(-2x+2)=-17(x+1)$

37. $2(6-2r)=-5(r-3)$

38. $-3(7p+2)-2(2-12p)=2(4+p)$

39. $-5(3-2z)+5(1-z)=6(2+z)$

Solve each equation and then check the solution in exercises 40-69.

40. $7x = 28$

41. $9x = 63$

42. $2a = -10$

43. $5k = -80$

44. $10t = -32$

45. $4s = -38$

46. $-8x = -96$

47. $-7x = -49$

48. $5r = 0$

49. $8x = 0$

50. $\frac{1}{3}y = -18$

51. $\frac{1}{7}p = -2$

52. $-y = 14$

53. $-t = 19$

54. $-x = -\frac{2}{9}$

55. $-m = -\frac{8}{7}$

56. $0.3t = 9$

57. $0.7x = 14$

58. $3x + 2x = 20$

59. $8x + 3x = 132$

60. $8m + 7m - 3m = 72$

61. $13r - 7r + 9r = 105$

62. $\frac{x}{5} = -7$

63. $\frac{k}{9} = -8$

64. $-\frac{3}{4}p = -9$

65. $-\frac{4}{9}y = -5$

66. $-\frac{2}{7}c = \frac{8}{17}$

67. $-\frac{7}{2}d = \frac{5}{6}$

68. $-1.4m = 21.14$

69. $-8.1a = -23.49$

For exercises 70-71, write an equation using the information given in the problem. Use x as the variable. Then solve the equation.

70. When a number is divided by –4, the result is 3. Find the number.

71. If twice a number is divided by 4, then result is 3. Find the number.

2.2   Solving Linear Equations

Procedure for Solving a Linear Equation

To solve linear equations, follow these steps.

Step 1   **Simplify each side separately.**   Use the distributive property to clear parentheses and combine terms, as needed.

Step 2   **Isolate the variable terms on one side.**   If necessary, use the addition property to get all variable terms on one side of the equation and all numbers on the other.

Step 3   **Isolate the variable.**   Use the multiplication property, if necessary, to get the equation in the form $x = $ a number.

Step 4   **Check.**   Check the solution by substituting into the *original* equation.

EXAMPLE 1

Solve $5(k - 2) - 2k = k - 12$.

Step 1   Before combining like terms, use the distributive property to simplify $5(k - 2)$.

$$5(k - 2) - 2k = k - 12$$

$$5 \cdot k - 5 \cdot 2 - 2k = k - 12 \qquad \text{Distributive property}$$

$$5k - 10 - 2k = k - 12$$

$$3k - 10 = k - 12 \qquad \text{Combine like terms.}$$

Step 2 $-k + 3k - 10 = -k + k - 12 \qquad$ Add $-k$.

$$2k - 10 = -12$$

$$2k - 10 + 10 = -12 + 10 \qquad \text{Add 10.}$$

$$2k = -2$$

Step 3 $\dfrac{2k}{2} = \dfrac{-2}{2} \qquad$ Divide by 2.

$$k = -1$$

Step 4   Check your answer by substituting -1 for k in the original equation. Remember to do the work inside the parentheses first.

$$5(k - 2) - 2k = k - 12$$

$$5(-1 - 2) - 2(-1) = -1 - 12 \qquad ? \qquad \text{Let } k = -1.$$

$$5(-3) - 2(-1) = -1 - 12 \qquad ?$$

$$-15 - 2(-1) = -1 - 12 \qquad ?$$

$$-15 - (-2) = -1 - 12 \qquad ?$$

$$-15 + 2 = -1 - 12 \qquad ?$$

$$-13 = -13 \qquad \text{True}$$

EXAMPLE 2

Solve $7a - (5 + 3a) = 2a + 8$.

Step 1   Simplify.

$$7a - (5 + 3a) = 2a + 8$$

$$7a - 5 - 3a = 2a + 8 \qquad \text{Distributive property}$$

$$4a - 5 = 2a + 8 \qquad \text{Combine like terms.}$$

Step 2 $-2a + 4a - 5 = -2a + 2a + 8 \qquad$ Add $-2a$.

$$2a - 5 = 8$$

$$2a - 5 + 5 = 8 + 5 \qquad \text{Add 5.}$$

$$2a = 13$$

Step 3 $\dfrac{2a}{2} = \dfrac{13}{2} \qquad$ Divide by 2.

$$a = \dfrac{13}{2}$$

Step 4   Check the solution in the original equation.

$$7a - (5 + 3a) = 2a + 8$$

$$7\left(\frac{13}{2}\right) - \left[5 + 3\left(\frac{13}{2}\right)\right] = 2\left(\frac{13}{2}\right) + 8 \quad ? \qquad \text{Let } a = \frac{13}{2}.$$

$$\frac{91}{2} - \left[5 + \frac{39}{2}\right] = 13 + 8 \qquad ?$$

$$\frac{91}{2} - \left[\frac{10}{2} + \frac{39}{2}\right] = 21 \qquad ?$$

$$\frac{91}{2} - \left(\frac{49}{2}\right) = 21 \qquad ?$$

$$\frac{42}{2} = 21 \qquad ?$$

$$21 = 21 \qquad\qquad\qquad \text{True}$$

Be very careful with signs when solving equations like the one in Example 2. When a subtraction sign appears immediately in front of a quantity in parentheses, such as in the expression

$$7a - (5 + 3a),$$

remember that the $-$ sign acts like a factor -1 and affects the sign of *every* term within the parentheses. Thus,

$$7a - (5 + 3a) = 7a + (-1)(5 + 3a) = 7a - 5 - 3a$$

$$\uparrow \quad \uparrow$$

Change to $-$ in *both* terms.

Equations with Fractions and Decimals as Coefficients: We can **clear** the equation (get integer coefficients) by using the multiplication property.

EXAMPLE 3

Solve $0.25t + 0.15(30 - 2t) = 0.21(20)$.

Step 1   Begin be clearing parentheses.

$$0.25t + 0.15(30 - 2t) = 0.21(20)$$

$$0.25t + 0.15(30) - 0.15(2t) = 0.21(20) \qquad \text{Distributive property}$$

To clear decimals, multiply both sides of the equation by 100, since the decimals are all hundredths. A number can be multiplied by 100 by moving the decimal point two places to the right.

$$0.25t + 0.15(30) - 0.15(2t) = 0.21(20)$$

$$25t + 15(30) - 15(2t) = 21(20) \qquad \text{Multiply by 100.}$$

$$25t + 450 - 30t = 420$$

$$-5t + 450 = 420 \qquad \text{Combine like terms.}$$

Step 2 $$-5t + 450 - 450 = 420 - 450 \qquad \text{Subtract 450.}$$

$$-5t = -30$$

Step 3 $$\frac{-5t}{-5} = \frac{-30}{-5} \qquad \text{Divide by } -5.$$

$$t = 6$$

Step 4   Check that $\{6\}$ is the solution by substituting into the original equation.

When clearing equations of fractions or decimals be sure to multiply *every* term on both sides of the equation by the least common denominator.

Self-Check 1

Solve the following equations.

1. $5(x - 4) - x = 2x - 6$ **2.** $7z - (5 + 2z) = 2z + 8$

3. $0.75t + 0.20(17 - t) = 0.05 \cdot 134$ **4.** $\dfrac{6}{5}x + 2(x - 6) = \dfrac{x}{5}$

Equations with No Solutions or Infinitely Many Solutions: Each equation that we have solved so far has had exactly one solution. As the next examples show, linear equations may also have no solutions or infinitely many solutions. (The four steps are not identified in these examples. See if you can identify them.)

EXAMPLE 4

Solve $7x - 35 = 7(x - 5)$.

$$7x - 35 = 7(x - 5)$$
$$7x - 35 = 7x - 35 \qquad \text{Distributive property}$$
$$-7x + 7x - 35 = -7x + 7x - 35 \qquad \text{Add } -7x.$$
$$-35 = -35$$
$$35 - 35 = 35 - 35 \qquad \text{Add 35.}$$
$$0 = 0$$

The variable has "disappeared." Since the last statement $(0 = 0)$ is *true, any* real number is a solution. (We could have predicted this from the line in the solution that says $7x - 35 = 7x - 35$, which is certainly true for *any* value of x.) Indicate the solution set as {all real numbers}.

EXAMPLE 5

Solve $7x + 2(x + 3) = 9x + 3$.

$$7x + 2(x + 3) = 9x + 3$$
$$7x + 2x + 6 = 9x + 3 \qquad \text{Distributive property}$$
$$9x + 6 = 9x + 3 \qquad \text{Combine terms.}$$
$$9x + 6 - 9x = 9x + 3 - 9x \qquad \text{Subtract } 9x \text{ from each side.}$$
$$6 = 3 \qquad \text{Combine terms.}$$

Again, the variable has disappeared, but this time a *false* statement $(6 = 3)$ results. When this happens, the equation has no solution. Its solution set is the **empty set**, or **null set**, symbolized $\varnothing$.

The chart below summarizes the solution sets of the three types of linear equations.

Final Equation	Number of Solutions	Solution Set
$x = $ a number	One	{a number}
A true statement with no variable, such as $0 = 0$.	Infinite	{all real numbers}
A false statement with no variable, such as $6 = 3$.	None	$\varnothing$

Self-Check 2

Solve the following equations.

1. $3x - 2 = 4(x - 1) - x - 2$ **2.** $3x - 2 = 4(x - 1) - (x - 2)$

3. $5x + 2(x - 7) = 3x + 4(x - 1)$

Translating Phrases into an Algebraic Expression: Often we are given a problem in which the sum of two quantities is a particular number, and we are asked to find the values of the two quantities. The next example shows how to express the unknown quantities in terms of a single variable.

EXAMPLE 6

Two numbers have a sum of 48. If one of the numbers is represented by k, find an expression for the other number.

First, suppose that the sum of two numbers is 48, and one of the numbers is 18. How would you find the other number? You would subtract 18 from 48 to get 30: $48 - 18 = 30$. So instead of using 18 as one of the numbers, use k as stated in the problem. The other number would be obtained in the same way. You must subtract k from 48. Therefore, an expression for the other number is $48 - k$.

Self-Check 3

Write the answer to each of the following as an algebraic expression.

1. Two numbers have a sum of 100. One of the numbers is n. Find the other number.

2. The product of two numbers is 144. One of the numbers is x. What is the other number?

3. Nadia is one year younger than Nicole. If a is the age of Nicole, what is the age of Nadia?

4. Amina has n nickels and d dollars in her piggy bank. Find the value of her money in cents.

Self-Check answers

1.1 $\{7\}$ **1.2** $\left\{\frac{13}{3}\right\}$ **1.3** $\{6\}$ **1.4** $\{4\}$

2.1 $\varnothing$ **2.2** $\{$all real numbers$\}$ **2.3** $\varnothing$

3.1 $100 - n$ **3.2** $\frac{144}{x}$ **3.3** $a - 1$ **3.4** $5n + 100d$

2.2 EXERCISES

1. In your own words, give the four steps to solve a linear equation. Use an example to demonstrate the steps.

Solve each equation and check your solution in exercises 2-18.

2. $3k + 5 = 23$ **3.** $4m + 12 = 20$ **4.** $14h - 7 = 12h + 7$

5. $-5x - 5 = -3x + 5$ **6.** $-2p + 5 = -(5p + 2)$ **7.** $6x + 2 = -(x - 9)$

8. $2(5x + 4) = -3(x - 1)$ **9.** $3(4m + 1) = 5(2m + 9)$ **10.** $8(2w + 5) = 3(6w + 10)$

11. $6(3x - 1) = 9(2x + 3)$

12. $8(3x + 7) = 4(6x - 1)$

13. $6(8x - 4) = 12(4x - 2)$

14. $3(6 - 2x) = 2(-3x + 9)$

15. $8r - 6r + 2 = 6r - 2r$

16. $10p - 5p + 6 = 8p - 4p$

17. $10x - 4(x + 9) - 6x = 0$

18. $6x - 9(x + 7) + 3x = 0$

19. Which one of the following linear equations does not have all real numbers as its solution?

(a) $7x = 6x + x$

(b) $3(x + 4) = 3x + 12$

(c) $\dfrac{1}{4}x = 0.25x$

(d) $7x = 2x$

20. Explain why an equation involving $\dfrac{1}{x - 5}$ cannot have 5 as a solution.

Solve each equation by clearing fractions or decimals and check your solution in exercises 21-32.

21. $\dfrac{2}{3}t - \dfrac{1}{6}t = t - \dfrac{5}{2}$

22. $-\dfrac{3}{5}r + r = \dfrac{1}{3}r + \dfrac{5}{3}$

23. $-\dfrac{1}{6}(x - 12) + \dfrac{1}{3}(x + 6) = x + 4$

24. $\dfrac{1}{8}(y + 10) + \dfrac{1}{4}(2y + 3) = y + 2$

25. $\dfrac{1}{4}k - \left(k + \dfrac{1}{3}\right) = \dfrac{1}{12}(k + 2)$

26. $-\dfrac{1}{6}q - \left(q - \dfrac{1}{4}\right) = \dfrac{1}{8}(q + 3)$

27. $0.25(10) + 0.15x = 0.05(60 + x)$

28. $0.15(20) + 0.30x = 0.25(30 + x)$

29. $1.15x + 0.15(70 - x) = 0.25(50)$

30. $0.20x + 0.70(1 - x) = 0.22(10)$

31. $0.09(10{,}000) + 0.02x = 0.07(10{,}000 + x)$

32. $0.03(5{,}000) - 0.03x = 0.015(7{,}000 + x)$

Solve each equation and check your solution in exercises 33-44.

33. $8(6x - 1) = 4(4x + 1) + 20$

34. $9(2k - 5) = 7(3k - 1) - 50$

35. $-(5y + 3) - (-4y - 8) = 5$

36. $-(9k - 8) - (-8k + 5) = 3$

37. $\dfrac{1}{2}(x + 5) + \dfrac{2}{3}(x + 3) = x + 7$

38. $\dfrac{1}{3}(x - 2) + \dfrac{1}{6}(x - 8) = x + 5$

39. $0.20(x + 70) + 0.30x = 24$

40. $0.30(x + 210) + 0.20(x + 300) = 45$

41. $6(x + 8) = 3(2x + 6) + 30$

42. $12(x + 3) = 4(3x + 8) + 4$

43. $9(v + 1) - v = 2(4v + 1) - 9$

44. $8(t - 4) + 10t = 6(3t + 1) - 10$

Write the answer to each problem as an algebraic expression in exercises 45-48.

45. Two numbers have a sum of 17. One of the numbers is q. Find the other number.

46. The product of two numbers is 21. One of the numbers is k. What is the other number?

47. Faiz is a years old. How old will he be in 14 years? How old was he 7 years ago?

48. Dean has r dimes. Find the value of the dimes in cents.

2.3   Problem Solving

The Six-Step Method for Solving an Applied Problem: We now begin to look at how algebra is used to solve applied problems. Some of the problems you will encounter may seem "contrived," and to some extent they are. But the skills you develop in solving simple problems will help you in solving more realistic problems in chemistry, physics, biology, business, and other fields.

There is no specific method that enables you to solve all kinds of applied problems, but the following six-step method is suggested.

The Six-Step Method for Solving an Applied Problem

Step 1   **Decide what you are asked to find.** Read the problem carefully. Choose a variable to represent the number you are asked to find. *Write down* what the variable represents.

Step 2   **Write down any other pertinent information.** If there are other unknown quantities, express them using the variable. Draw figures or diagrams and use charts, if they apply.

Step 3   **Write an equation.** Translate the problem into an equation.

Step 4   **Solve the equation.** Use the properties to solve the equation.

Step 5   **Answer the question(s) posed.** Make sure you answer the question in the problem. In some cases, you need more than the solution of the equation.

Step 6   **Check.** Check your solution using the original words of the problem. Be sure your answer makes sense.

The third step is often the hardest. To translate the problem into an equation, write the given phrases as mathematical expressions. Translate any words that mean *equal* or *same as* =. The = sign leads to an equation to be solved.

EXAMPLE 1

The product of 5, and a number decreased by 6, is 200. Find the number.

Step 1   Read the problem carefully. Decide what you are being told to find, and then choose a variable to represent the unknown quantity. In this problem, we are told to find a number, so we write

$$\text{Let } x = \text{ the number.}$$

Step 2   There are no other unknown quantities to find.

Step 3   Translate the problem into an equation.

The product of 5,	times	and	a number	decreased by	6,	is	200.
↓	↓		↓	↓	↓	↓	↓
5	·	(	x	−	6)	=	200

Because of the commas in the given problem, writing the equation as $5x - 6 = 200$ is incorrect. The equation $5x - 6 = 200$ corresponds to the statement, "The product of 5 and a number, decreased by 6, is 200."

Step 4   Solve the equation.

$$5 \cdot (x - 6) = 200$$

$5x - 30 = 200$	Distributive property
$5x = 230$	Add 30 to both sides.
$x = 46$	Divide by 5.

Step 5   The number is 46.

Step 6   Check the solution by using the original words of the problem. When 46 is decreased by 6, we get $46 - 6 = 40$. If 5 is multiplied by 40, we get 200, as the problem required. The answer, 46, is correct.

Solve Problems Involving Sums of Quantities: A common type of problem in algebra involves finding two quantities when the sum of the quantities is known. In Example 6 of the previous section, we prepared for this type of problem by writing mathematical expressions for two related unknown quantities.

In general, to solve such problems, choose a variable to represent one of the unknowns and then represent the other quantity in terms of the same variable, using information obtained in the problem. Then write an equation based on the words of the problem. The next example illustrates these ideas.

EXAMPLE 2

In the 2000 Olympics, U.S. contestants won 16 more gold than silver medals. They won a total of 64 gold and silver medals. Find the number of each type of medal won. *(Source: United States Olympic Committee.)*

Step 1   Let x the number of silver medals.

Step 2   Let $x + 16$ the number of gold medals.

Step 3   Now write an equation.

The total	is	the number of silver	plus	the number of gold.
↓	↓	↓	↓	↓
64	=	x	+	$(x + 16)$

Step 4   Solve the equation.

Step 5   Because x represents the number of silver medals, the U.S. won 24 silver medals. Because $x + 16$ represents the number of gold medals, the U.S. won $24 + 16 = 40$ gold medals.

Step 6   Since there were 40 gold and 24 silver medals, the total number of medals was $40 + 24 = 64$. Because $40 - 24 = 16$, there were 16 more gold medals than silver medals. This information agrees with what is given in the problem, so the answers check.

The problem in Example 2 could also have been solved by letting x represent the number of gold medals. Then $x - 16$ would represent the number of silver medals. The equation would then be.

$$64 = x + (x - 16)$$

The solution of this equation is 40, which is the number of gold medals. The number of silver medals would then be $40 - 16 = 24$. The answers are the same, whichever approach is used.

Self-Check 1

Solve the following problems.

1. The product of a number and 5, increased by 2 is 37. Find the number.

2. The product of a number, and 5 increased by 2 is 42. Find the number.

3. Some lumber is cut into two pieces such that one piece is two inches longer than the other. If the lumber was 24 inches long, find the length of each piece.

4. Some lumber is cut unto two pieces such that one piece is twice as long as the other. If the lumber is 24 inches long, find the length of each piece.

Three More Common Applied Problems: Mixture problems commonly can be analyzed in a six-step approach.

EXAMPLE 3

A lawn trimmer uses a mixture of gasoline and oil. For each ounce of oil the mixture contains 16 ounces of gasoline. If the tank holds 68 ounces of the mixture, how many ounces of oil and how many ounces of gasoline does it require when it is full?

Step 1   Let x the number of ounces of oil required when full.

Step 2   Let $16x =$ the number of ounces of gasoline required when full.

	Amount of gasoline		Amount of oil		Total amount in tank
	↓		↓		↓
Step 3	$16x$	$+$	x	$=$	68

Step 4 $17x = 68$ Combine terms.
                         $x = 4$ Divide by 17.

Step 5   The trimmer requires 4 ounces of oil and $16(4) = 64$ ounces of gasoline when full.

Step 6   Since $4 + 64 = 68$, and 64 is 16 times 4, the answers check.

Sometimes it is necessary to find three unknown quantities in an applied problem. Frequently the three unknowns are compared in *pairs*. When this happens, it is usually easiest to let the variable represent the unknown found in both pairs. The next example illustrates this strategy.

EXAMPLE 4

A woodworking project requires three pieces of wood. The longest piece must be twice the length of the middle-sized piece, and the shortest piece must be 10 inches shorter than the middle-sized piece. Bara Baker has a board 110 inches long that she wishes to use. How long can each piece be?

Steps 1 and *2*   Since the middle-sized piece appears in both pairs of comparisons, let x represent the length of the middle-sized piece. We have

$$x = \text{the length of the middle - sized piece}$$
$$2x = \text{the length of the longest piece}$$
$$x - 10 = \text{the length of the shortest piece.}$$

Figure 1

	Longest ↓		Middle sized ↓		Shortest ↓		Total length ↓
Step 3	$2x$	$+$	x	$+$	$(x - 10)$	$=$	110

Step 4

$$4x - 10 = 110 \qquad \text{Combine terms.}$$
$$4x - 10 + 10 = 110 + 10 \qquad \text{Add 10 to each side.}$$
$$4x = 120 \qquad \text{Combine terms.}$$
$$x = 30 \qquad \text{Divide by 4 on each side.}$$

Step 5   The middle-sized piece is 30 inches long, the longest piece is $2(30) = 60$ inches long, and the shortest piece is $30 - 10 = 20$ inches long.

Step 6   Check to see that the sum of the lengths is 110 inches, and that all conditions of the problem are satisfied.

The next example deals with concepts from geometry. An angle can be measured by a unit called the **degree** (°). Two angles whose sum is 90° are said to be **complementary**, or complements of each other. Two angles whose sum is 180° are said to be **supplementary**, or supplements of each other. See Figure 2. If x represents the degree measure of an angle, then

$$90 - x = \text{the degree measure of its complement, and}$$
$$180 - x = \text{the degree measure of its supplement.}$$

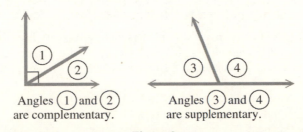

Angles ① and ②
are complementary.

Angles ③ and ④
are supplementary.

Figure 2

EXAMPLE 5

Find the measure of an angle whose supplement is 50° more than three times its complement.

Step 1   Let $180 - x =$ the degree measure of its supplement.

Step 2   Then $90 - x =$ the degree measure of its complement;
                        $x =$ the degree measure of the angle.

Step 3

Supplement	is	50	more than	three	times	its complement
↓	↓	↓	↓	↓	↓	↓
$180 - x$	$=$	50	$+$	3	$\cdot$	$(90 - x)$

Step 4   Solve the equation.

$$180 - x = 50 + 270 - 3x \qquad \text{Distributive property}$$
$$180 - x = 320 - 3x \qquad \text{Combine terms.}$$
$$180 - x + 3x = 320 - 3x + 3x \qquad \text{Add } 3x.$$
$$180 + 2x = 320 \qquad \text{Combine terms.}$$
$$180 + 2x - 180 = 320 - 180 \qquad \text{Subtract 180.}$$
$$2x = 140 \qquad \text{Combine terms.}$$
$$x = 70 \qquad \text{Divide by 2.}$$

Step 5   The measure of the angle is 70°

Step 6   The complement of 70° is 20° and the supplement of 70° is 110°. 110° is equal to 50° more than three times 20° ($110 = 50 + 3(20)$ is true); therefore, the answer is correct

---

Self-Check 2

Solve the following problems.

1. A lemon drink is a mixture of lemon juice and water. For each ounce of pure lemon juice the drink contains 24 ounces of water. If a pitcher holds 100 ounces of drink, how many ounces of pure lemon juice and water does it require to be filled?

2. A teacher is giving treats to her students who are all the same age. The student that got the highest mark got twice as many treats as her age. The student that received the lowest mark received 5 less treats than her age. The other student received the same number of treats as his age. If the teacher gave 43 treats to these three students, what are the ages of her students?

3. Find the measure of an angle whose supplement is 26° less than three times its complement.

Consecutive Integer Problems: Two integers that differ by 1 are called **consecutive integers**. For example, 3 and 4, 6 and 7, and −2 and −1 are pairs of consecutive integers. In general, if x represents an integer, $x + 1$ represents the next larger consecutive integer.

Consecutive *even* integers, such as 8 and 10, differ by 2. Similarly, consecutive *odd* integers, such as 9 and 11, also differ by two. In general, if x represents an even integer $x + 2$ represents the next larger consecutive even integer. The same holds true for odd integers; that is, if x is an odd integer, $x + 2$ is the next larger odd integer.

EXAMPLE 6

Two pages that face each other in a book have 685 as the sum of their page numbers. What are the page numbers?

Because the two pages face each other, they must have page numbers that are consecutive integers.

Let $\quad x =$ the smaller page number.
Then $\quad x + 1 =$ the larger page number.

Because the sum of the page numbers is 685, the equation is

$$x + (x + 1) = 685.$$

Solve the equation.

$$x + (x + 1) = 685$$
$$2x + 1 = 685 \quad \text{Combine like terms.}$$
$$2x = 684 \quad \text{Subtract 1.}$$
$$x = 342 \quad \text{Divide by 2.}$$

The smaller page number is 342 and the larger page number is $343 + 1 = 343$. The sum of 342 and 343 is 685. Our answer is correct.

EXAMPLE 7

If the smaller of two consecutive odd integers is tripled, the result is 20 more than the larger of the two integers. Find the two integers.

Let x be the smaller integer. Since the two numbers are consecutive *odd* integers, then $x + 2$ is the larger. Now write an equation.

If the smaller is tripled	the result is	20	more than	the larger.
↓	↓	↓	↓	↓
$3x$	$=$	20	+	$x + 2$

Solve the equation.

$$3x = 20 + x + 2$$
$$3x = 22 + x$$
$$3x + (-x) = 22 + x + (-x)$$
$$2x = 22$$
$$x = 11$$

The first integer is 11 and the second is $11 + 2 = 13$. To check our answer we see that when 11 is tripled, we get 33, which is 20 more than the larger odd integer, 13. Our answer is correct.

Self-Check 3

Solve the following problems.

1. The sum of two consecutive even integers is 126. What is the smaller of the two integers?

2. The sum of three consecutive odd integers is 165. What is the largest of the three integers?

3. If the smaller number of two pages facing each other is tripled, the result is 98 more than the larger number being doubled. What is the page number on the right?

Self-Check Answers
1.1 7   **1.2** 6   **1.3** 11 in. and 13 in.   **1.4** 8 in. and 16 in.
2.1 4 ounces of pure lemon juice and 96 ounces of water
2.2 They are each 12 years old.   **2.3** 32°
3.1 62   **3.2** 57   **3.3** 101

2.3 EXERCISES

1. Which one of the following would not be a reasonable answer in an applied problem that requires finding the number of marbles in a bag?
    (a) 7   (b) 0   (c) $3\frac{1}{2}$ (d) 71

2. Which one of the following would not be a reasonable answer in an applied problem that requires finding the age of a dog?
    (a) 9   (b) –4   (c) $3\frac{1}{2}$ (d) $6\frac{1}{4}$

3. Explain in your own words the six-step method for solving applied problems.

4. List some words that will translate as "=" in an applied problem.

Solve each problem in exercises 5-8. Use the six-step method.

5. If 2 is added to a number and this sum is doubled, the result is 11 more than the number. Find the number.

6. If 4 is subtracted from a number and this difference is tripled, the result is 6 more than the number. Find the number.

7. If 4 is added to twice a number and this sum is multiplied by 3, the result is the same as if the number is multiplied by 5 and 12 is added to the product. What is the number?

8. The sum of two times a number and 8 more than the number is the same as the difference between 23 and twice the number. What is the number?

Solve each problem in exercises 9-22. Use the six-step method.

9. Niagara Falls lies on the border between New York State and Ontario, Canada. Separated by Goat Island, the falls are divided into two main parts, the American Falls and Horseshoe Falls. Horseshoe Falls is higher by 2 meters. If the combined height is 114 meters, what is the height of each fall? (*Source:* Encyclopedia Britannica)

10. The total number of Democrats and Republicans in the U.S. House of Representatives in 2001 for the 107th Congress was 431. There were 9 fewer Democrats than Republicans. How many members of each party were there? (*Source:* U.S. House of Representatives, Office of the Clerk)

11. During the 2000 Olympics in Sydney, Australia there were 2513 more men than women competing. The total number of competitors was 10,651. How many men and how many women competed? (*Source:* The Olympic Museum)

12. The Baltimore Ravens and the New York Giants were the teams that competed in the Superbowl in Tampa, Florida on January 28, 2001. Baltimore won the game over New York by 27 points. If their combined total was 41 points, what was the final score for each team? (*Source*: CBS.Sportsline.)

13. Ali Baker has a strip of paper 36 inches long. He wants to cut it into two pieces so that one piece will be 10 inches shorter than the other. How long should the two pieces be?

14. In her job as a math tutor, Katerina Kucera works a $6\frac{1}{2}$-hour day. She tutors calculus, trigonometry, and algebra. One day she tutored algebra twice as long as calculus, and she tutored trigonometry $\frac{1}{2}$ hour longer than calculus. How many hours did she spend on each subject?

15. In 2000, there were 120 seats on the State of Florida's House of Representatives. If there were 34 more Republicans than Democrats, how many were there from each party? (*Source: Official Site of the Florida Legislature*)

16. North Carolina and Virginia are bordering U.S. states on the east coast. Virginia's area is 11,052 square miles less than North Carolina's. If together their area is 96,590 square miles then what is the area of each? (*Source: www.50states.com*)

17. Earth, Neptune and Mars together have 11 satellites (moons). If Earth only has the one moon and Neptune has four times as many moons as Mars, how many satellites does each planet have? (*Source*: Students for the Exploration and Development of Space (www.seds.org))

18. The sum of the measures of the angles of any triangle is 180°. In triangle *ABC*, angles *A* and *B* have the same measure, while the measure of angle *C* is 48° larger than each of *A* and *B*. What are the three angles?

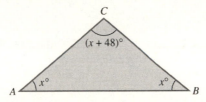

19. (See Exercise 18.) In triangle *ABC*, the measure of angle *A* is 117° more than the measure of angle *B*. The measure of angle *B* is the same as the measure of angle *C*. Find the measure of each angle.

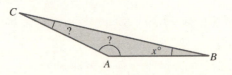

20. A pharmacist found that at the end of the day she had twice as many prescriptions for antibiotics as she did for tranquilizers. She had 51 prescriptions altogether for these two types of drugs. How many did she have for tranquilizers?

21. A mixture of nuts contains only peanuts and cashews. For every ounce of cashews there are 6 ounces of peanuts. If the mixture contains a total of 42 ounces, how many ounces of each type of nut does the mixture contain?

22. An insecticide contains 79 centigrams of inert ingredient for every 1 centigram of active ingredient. If a quantity of the insecticide weighs 370 centigrams, how much of each type of ingredient does it contain?

Use the concepts of this section to answer each question in exercises 23-32.

23. If the sum of two numbers is *p,* and one of the numbers is *q,* how can you express the other number?

24. Is there an angle whose supplement is equal to its complement? If so, what is the measure of the angle?

25. Is there an angle that is equal to its supplement? Is there an angle that is equal to its complement? If the answer is yes to either question, give the measure of the angle.

26. Express three consecutive even integers, all in terms of *x,* if *x* represents the middle of the three.

27. Find the measure of an angle whose supplement measures 7 times the measure of its complement.

28. Find the measure of an angle whose supplement measures 3 times the measure of its complement.

29. Find the measure of an angle, if its supplement measures 16° less than three times its complement.

30. Find the measure of an angle, if its supplement measures 65° more than twice its complement.

31. Find the measure of an angle such that the sum of the measures of its complement and its supplement is 120°.

32. Find the measure of an angle such that the difference between the measures of its supplement and three times its complement is 16°.

Solve each problem in exercises 33-41.

33. The sum of two consecutive integers is 127. Find the integers.

34. The sum of two consecutive integers is 139. Find the integers.

35. Find two consecutive even integers such that the smaller added to three times the larger gives a sum of 54.

36. Find two consecutive odd integers such that twice the larger is 35 more than the smaller.

37. Two pages that are back-to-back in this book have 223 as the sum of their page numbers. What are the page numbers?

38. If the sum of three consecutive even integers is 78, what is the smallest even integer?

39. When the smaller of two consecutive integers is added to three times the larger, the result is 71. Find the integers.

40. If five times the smaller of two consecutive integers is added to three times the larger, the result is 235. Find the integers.

41. If the first and third of three consecutive odd integers are added, the result is 22 less than four times the second integer. Find the integers.

2.4   Formulas and Applications from Geometry

Many applied problems can be solved with formulas. There are formulas for geometric figures such as squares and circles, for distance, for money earned on bank savings, and for converting English measurements to metric measurements.

If you are given the values of all but one of the variables in a formula, you can find the value of the remaining variable by using the methods for solving equations.

EXAMPLE 1

The formula for the area of a rectangle with length L and width W is $A = LW$. Given that $A = 57$ and $L = 10$, solve for W. Substitute into the formula.

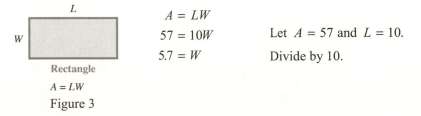

$$A = LW$$
$$57 = 10W \qquad \text{Let } A = 57 \text{ and } L = 10.$$
$$5.7 = W \qquad \text{Divide by 10.}$$

Rectangle
$A = LW$
Figure 3

Check that the width of the rectangle is 5.7.

Applications: Perimeter, Circumference, and Area: For applications involving geometric figures, a drawing or diagram is often helpful.

Recall that the **perimeter** of a two-dimensional figure is the distance around the figure, or the sum of the lengths of its sides. Similarly, the circumference of a ball (or sphere) is the greatest distance around the ball. (See Figure 4.)

EXAMPLE 2

A professional basketball has a circumference of at most 78 centimeters. See Figure 4. What is the radius of the circular cross section through the center of the ball?

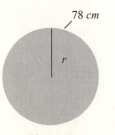

78 cm

r

Figure 4

We want to find the radius shown, so let $r =$ the radius in centimeters.

The formula for the circumference of a circle is $C = 2\pi r$.

To find the radius r, we substitute 78 for C. We will use approximate value of 3.14 for π.

$$78 = 2(3.14)r \qquad C = 78,\ \pi \approx 3.14$$

Solve the equation.

$$78 = 6.28r \quad \text{Multiply.}$$
$$r = 12.42 \quad \text{Divide by 6.28; round.}$$

The radius of the indicated cross section is 12.42 centimeters.

Check. If $r = 12.42$ and $\pi \approx 3.14$, the circumference is $2(3.14)(12.42)$ or about 78, as required.

The **area** of a geometric figure is a measure of the surface covered by the figure. Example 3 shows an application of area. We use the six steps of problem solving.

EXAMPLE 3

The area of a triangular sail of a sailboat is 143 square feet. The base of the sail is 13 feet. Find the height of the sail.

Step 1   Since we must find the height of the triangular sail,

$$\text{let } h = \text{the height of the sail in feet.}$$

Step 2   See Figure 5.

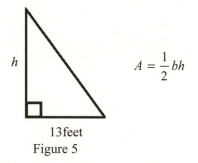

$$A = \frac{1}{2}bh$$

13 feet

Figure 5

Step 3   The formula for the area of a triangle is $A = \frac{1}{2}bh$ where A is the area, b is the base, and h is the height. Using the information given in the problem, substitute 143 for A and 13 for b in the formula.

$$A = \frac{1}{2}bh$$

$$143 = \frac{1}{2}(13)h \qquad A = 143,\ b = 13$$

Step 4   Solve the equation.

$$143 = \frac{13}{2}h$$

$$\frac{2}{13}(143) = \frac{2}{13}\left(\frac{13}{2}h\right) \qquad \text{Multiply by } \tfrac{2}{13}.$$

$$22 = h \qquad\qquad\qquad \text{Simplify.}$$

Step 5   The height of the sail is 22 feet.

Step 6   Check to see that the values $A = 143$, $b = 13$, and $h = 22$ satisfy the formula for the area of a triangle.

Self-Check 1

Solve the following problems.

1. A rectangle has a area of 72 square inches. If the width of the rectangle is 9 inches, what is its length?

2. A rectangle has a perimeter of 66 centimeters. Its length is 13 centimeters. What is it's width? $(P = 2L + 2W)$

3. If a circle has a circumference of 10 meters, what is it's radius in meters rounded to two decimal places? (Use $\pi \approx 3.14$)

4. The area of a triangular piece of cloth is 21 square feet. What is the base of the cloth if the height is 6 feet?

Solve Problems about Angle Measures: Refer to Figure 6, which shows two intersecting lines forming angles that are numbered ①, ②, ③, and ④. Angles ① and ③ lie "opposite" each other. They are called **vertical angles**. Another pair of vertical angles are ② and ④. In geometry, it is shown that vertical angles have equal measures.

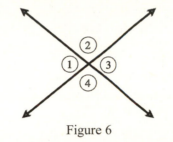

Figure 6

Now look at angles ① and ②. When their measures are added, we get 180°, the measure of a **straight angle**, so the angles are supplements. There are three other such pairs of angles: ② and ③, ③ and ④, ① and ④.

EXAMPLE 4

Find the measure of each marked angle in Figure 7.

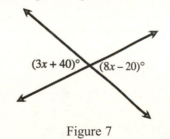

Figure 7

Since the marked angles are vertical angles, they have the same measures. Set $3x + 40$ equal to $8x - 20$ and solve.

$$3x + 40 = 8x - 20$$
$$-3x + 3x + 40 = -3x + 8x - 20 \quad \text{Add } -3x.$$
$$40 = 5x - 20$$
$$40 + 20 = 5x - 20 + 20 \quad \text{Add } 20.$$
$$60 = 5x$$
$$12 = x \quad \text{Divide by 5.}$$

Now we must find the required angle measures. Since $x = 12$, one angle has measure $3(12) + 40 = 36 + 40 = 76$ degrees. The other has the same measure, since $8(12) - 20 = 96 - 20 = 76$ as well. Each angle measures $76°$.

EXAMPLE 5

Find the measure of each marked angle in Figure 8.

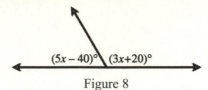

Figure 8

The measures of the marked angles must add to $180°$ since together they form a straight angle. The equation to solve is

$$(5x - 40) + (3x + 20) = 180.$$

$$8x - 20 = 180 \qquad \text{Combine terms.}$$

$$8x - 20 + 20 = 180 + 20 \quad \text{Add 20.}$$

$$8x = 200$$

$$x = 25 \qquad \text{Divide by 8.}$$

To find the measures of the angles, substitute 25 for x in the two expressions.

$$5x - 40 = 5(25) - 40 = 125 - 40 = 85$$

$$3x + 20 = 3(25) + 20 = 75 + 20 = 95$$

The two angle measures are $85°$ and $95°$, which are supplementary.

In Example 5, the answer is not the value of x. Remember to substitute the value of the variable into the expression for each angle you are asked to find.

Self-Check 2

Find the measures of each of the angles indicated.

1. 2.

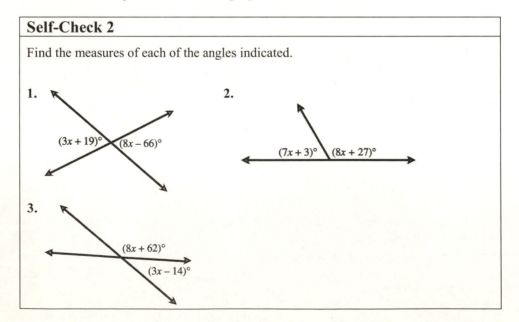

(3x + 19)° (8x − 66)°

(7x + 3)° (8x + 27)°

3.

(8x + 62)°

(3x − 14)°

Solve a Formula for a Specified Variable: Sometimes it is necessary to solve a number of problems that use the same formula. For example, a surveying class might need to solve several problems that involve the formula for the area of a rectangle, $A = LW$. Suppose that in each problem the area (A) and the length (L) of a rectangle are given and the width (W) must be found. Rather than solving for W each time the formula is used, it would be simpler to *rewrite the formula* so that it is solved for W. This process is called **solving for a specified variable**. Solving a formula for a specified variable requires the same steps used earlier to solve equations with just one variable.

The formula for converting temperatures in degrees Celsius to degrees Fahrenheit is

$$F = \frac{9}{5}C + 32.$$

The next example shows how to solve this formula for C.

EXAMPLE 6

Solve $F = \dfrac{9}{5}C + 32$ for C.

To clear the fraction, multiply both sides by 5.

$$F = \frac{9}{5}C + 32 \qquad\qquad \text{Solve for } C.$$

$$5F = 5\left(\frac{9}{5}C + 32\right) \qquad\qquad \text{Multiply by 5.}$$

$$5F = 9C + 160$$

Now, to get $9C$ alone on one side, subtract 160 from both sides. Then divide both sides by 9.

$$5F - 160 = 9C + 160 - 160 \qquad \text{Subtract 160.}$$

$$5F - 160 = 9C$$

$$\frac{5F - 160}{9} = C \qquad\qquad \text{Divide by 9.}$$

This result can be written in a different form using the distributive property to rewrite the numerator as $5(F - 32)$. Then we have

$$C = \frac{5(F - 32)}{9} = \frac{5}{9}(F - 32).$$

When solving a formula for a specified variable, treat that variable as if it were the *only* variable in the equation, and treat all others as if they were numbers. Then use the method of solving equations described earlier to solve for the specified variable.

Self-Check Answers

1.1 8 inches	**1.2** 20 centimeters	**1.3** 1.59 meters	**1.4** 7 feet
2.1 each is 70°	**2.2** 73° and 107°	**2.3** 22° and 158°	

2.4 EXERCISES

1. In your own words, explain what is meant by the *perimeter* of a geometric figure.

2. In your own words, explain what the *area* of a geometric figure means

In exercises 3-18, a formula is given along with the values of all but one of the variables in the formula. Find the value of the variable that is not given.

3. $P = 2L + 2W$ (perimeter of a rectangle); $L = 8, W = 2$

4. $P = 2L + 2W$; $L = 10, W = 13$

5. $P = 4s$ (perimeter of a square); $s = 8$

6. $P = 4s$; $s = 7$

7. $A = \frac{1}{2}bh$ (area of a triangle); $b = 15, h = 4$

8. $A = \frac{1}{2}bh$; $b = 6, h = 10$

9. $d = rt$ (distance formula); $d = 120, t = 4$

10. $d = rt$; $d = 135, r = 45$

11. $I = prt$ (simple interest); $p = 10000, r = 0.05, t = 6$

12. $I = prt$; $p = 1000, r = 0.045, t = 3$

13. $A = \frac{1}{2}h(b + B)$ (area of a trapezoid); $h = 5, b = 10, B = 8$

14. $A = \frac{1}{2}h(b + B)$; $h = 7, b = 29, B = 11$

15. $C = 2\pi r$ (circumference of a circle); $C = 15.7,\ \pi = 3.14$

16. $C = 2\pi r$; $C = 19.8448,\ \pi = 3.14$

17. $A = \pi r^2$ (area of a circle); $r = 8,\ \pi = 3.14$

18. $A = \pi r^2$; $r = 10,\ \pi = 3.14$

19. If a formula contains exactly five variables, how many values would you need to be given in order to find the value of any one variable?

20. The formula for changing Celsius to Fahrenheit given in Example 6 is $F = \frac{9}{5}C + 32$. Sometimes it is seen as $F = \frac{9C}{5} + 32$. These are both correct. Why is it true that $\frac{9}{5}C$ is equal to $\frac{9C}{5}$?

Find the measure of each marked angle in exercises 21-24.

21.

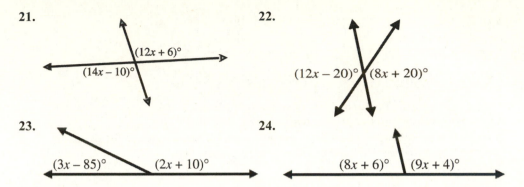

22.

23.

24.

Solve each formula for the specified variable in exercises 25-38.

25. $d = rt$ for r

26. $I = prt$ for r

27. $I = prt$ for p

28. $V = LWH$ for H

29. $P = a + b + c$ for b

30. $P = a + b + c$ for a

31. $y = mx + b$ for m

32. $Ax + By = C$ for y

33. $A = \dfrac{1}{2}bh$ for b

34. $A = \dfrac{1}{2}bh$ for h

35. $A = p + prt$ for r

36. $P = 2L + 2W$ for W

37. $V = \pi r^2 h$ for h

38. $V = \dfrac{1}{3}\pi r^2 h$ for h

CH 2 | Summary

KEY TERMS
2.1 linear equation in one variable   solution set
equivalent equations

2.2 empty (null set)

2.3 degree   complementary angles
supplementary angles   consecutive integers

2.4 perimeter   circumference
area   vertical angles
straight angles

CH 2 | Quick Review

2.1 THE ADDITION AND MULTIPLICATION PROPERTIES OF EQUALITIES
The same expression may be added to (or subtracted from) each side of an equation without changing the solution.

Each side of an equation may be multiplied (or divided) by the same nonzero expression without changing the solution.

2.2 SOLVING LINEAR EQUATIONS

Step 1   Clear parentheses and combine like terms to simplify each side.
Step 2   Get the variable term on one side, a number on the other.
Step 3   Get the equation into the form $x = $ a number.
Step 4   Check by substituting the result into the original equation.

2.3 SOME APPLICATIONS OF LINEAR EQUATIONS

Solving an Applied Problem Using the Six-Step Method
Step 1   Choose a variable to represent the unknown.
Step 2   Determine expressions for any other unknown quantities, using the variable. Draw figures or diagrams and use charts if they apply.
Step 3   Write an equation.
Step 4   Solve the equation.
Step 5   Answer the question(s) asked in the problem.
Step 6   Check your solution by using the original words of the problem. Be sure that the answer is appropriate and makes sense.

2.4 FORMULAS AND APPLICATIONS FROM GEOMETRY

To find the values of one of the variables in a formula given values for the others, substitute the known values into the formula.

To solve a formula for one of the variables, isolate that variable by treating the other variables as numbers and using the steps for solving equations.

CH 2 | Review Exercises

Solve each equation in exercises 1-18.

1. $m - 7 = 4$

2. $y + 5 = -8$

3. $4k + 5 = 3k + 12$

4. $6k = 5k + \dfrac{3}{4}$

5. $(8r - 5) - (7r + 2) = 10$

6. $4(3y - 5) = 5 + 11y$

7. $8k = 72$

8. $9r = -36$

9. $3p - 5p + 9p = 21$

10. $\dfrac{m}{14} = -1$

11. $\dfrac{7}{4}k = 4$

12. $11m + 6 = 72$

13. $2(2x + 5) - (x + 6) = x - 20$

14. $6x + 9 - (9x - 3) = 3x - 12$

15. $\dfrac{5}{4}r - \dfrac{r}{2} = \dfrac{3r}{4}$

16. $0.05(x + 900) + 0.45x = 50$

17. $4x - (-3x + 1) = 6(x - 9) + x$

18. $3(y - 4) - 5(y + 1) = -2(y + 10) + 3$

Use the six-step method to solve each problem in exercises 19-23.

19. Soccer is the world's most popular sport, with over 20 million participants. The number of teams participating in the 1998 World Cup when it was hosted in France was 8 more than when the World Cup was hosted in the U.S. in 1994. The number of teams in 1990 when the World Cup was hosted in Italy was the same as when it was hosted in the U.S. How many teams participated in 1994 if there was a total of 80 teams participating in those three years? (*Source: World Cup Archive*)

20. According to the 2000 Census, New York has a resident population greater than Florida by 2,994,079 residents. If the combined resident population is 34,958,835, how many resident live in each state?

21. The land area of Georgia is 3766 square miles greater than the area of Florida Together, the areas total 112,072 square miles. What is the area of each of the two states?

22. The two highest mountains in California are Mount Whitney and Mount Williamson. The sum of the two heights is 28,869 feet. If Mount Whitney is 119 feet higher than Mount Williamson, then what is the height of each mountain? (*Source: World Almanac and Book of Facts.*)

23. The supplement of an angle measures 6 times the measure of its complement. What is the measure of the angle?

A formula is given along with the values for all but one of the variables. Find the (approximate) value of the variable that is not given in exercises 24-27.

24. $A = \dfrac{1}{2}bh$; $A = 20$, $b = 10$ 25. $A = \dfrac{1}{2}h(b+B)$; $b = 6$, $B = 8, h = 10$

26. $C = 2\pi r$; $C = 14.13$, $\pi \approx 3.14$ 27. $V = \dfrac{4}{3}\pi r^3$; $r = 2$, $\pi \approx 3.14$

Solve the formula for each specified variable in exercises 28-29.

28. $A = LW$ for L 29. $V = \dfrac{1}{2}h(b+B)$ for h

Find the measure of each angle in exercises 30-31.

30. 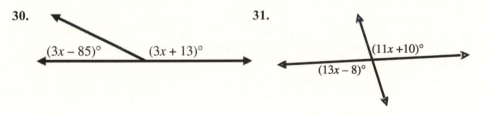   31.

$(3x - 85)°$ $(3x + 13)°$

$(11x + 10)°$

$(13x - 8)°$

CH 2 | Test

Solve each equation in exercises 1-5.

1. $7x + 5 = 10x + 26$ 2. $-\dfrac{5}{6}x = -20$

3. $8 - 2(m - 1) = -5m + 3(m + 11)$ 4. $0.05(x + 30) + 0.21(x - 10) = 2$

5. $-3(4x + 12) = -6(2x + 6)$

Solve the following problem.

6. According to the 2000 Census, the three states with the smallest population are Alaska, Vermont and Wyoming. Their populations total 1,729,541 residents. Alaska is populated by 18,105 more residents than Vermont. Wyoming has 115,045 less residents than Vermont. What is resident population of each state?

7. The formula for the perimeter of a rectangle is $P = 2L + 2W$.
   (a) Solve for W.
   (b) If $P = 76$ and $L = 25$, find the value of W.

Find the measure of each marked angle.

8.

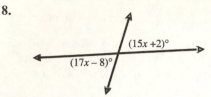

9. Find the measure of an angle if its supplement measures 12° more than four times its complement.

Solve each equation and check the solution in exercises 10-13.

10. $3r - 4 = 7r$

11. $4 - 4(a + 4) = 2(a + 3) - 12$

12. $\dfrac{3}{4}y + \dfrac{3}{5}y = -27$

13. $\dfrac{3x + 5}{7} = \dfrac{x - 9}{3}$

Solve each formula for the indicated variable in exercises 14-15.

14. $2x + 3y = 12$ for y

15. $A = P(1 + ni)$ for n

Solve the following problems in exercises 16-20.

16. In one day Scott Hochwald received 33 telephone calls. His wife called triple the number of times than his daughter, while his son called 2 less than the number of times his wife called. How many calls did he receive from his wife?

17. In a mixture of concrete, there are 3 pounds of cement mix for every 1 pound of gravel. If the mixture contains a total of 160 pounds of these two ingredients, how many pounds of gravel are there?

18. If the product of two numbers is r, and one of the numbers is s $(s \neq 0)$, how can you express the other number?

19. If x represents an integer, how can you express the next smaller consecutive integer in terms of x?

20. If 10 is subtracted from the largest of three consecutive odd integers, with this result multiplied by 2, the answer is 27 less than the sum of the first and twice the second of the integers. Find the integers.

CHAPTER 3: POLYNOMIAL AND EXPONENTS

| 3.1 | Integer Exponents and Scientific Notation |

In Chapter 1, we used exponents to write products of repeated factors. Now, in this section, we give further definitions and rules of exponents.

The Product Rule for Exponents: There are several useful rules that simplify work with exponents. For example, the product $2^4 \cdot 2^3$ can be simplified to 2^7 as follows.

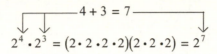

$$2^4 \cdot 2^3 = (2 \cdot 2 \cdot 2 \cdot 2)(2 \cdot 2 \cdot 2) = 2^7$$

To multiply exponential expressions with the same base, keep the same base and add the exponents:

Product Rule for Exponents

If m and n are natural numbers and a is any real number, then

$$a^m \cdot a^n = a^{m+n}.$$

EXAMPLE 1

(a) $3^6 \cdot 3^7 = 3^{6+7} = 3^{13}$

(b) $5^4 \cdot 5 = 5^4 \cdot 5^1 = 5^{4+1} = 5^5$

(c) $y^2 \cdot y^3 \cdot y^4 = y^{2+3+4} = y^9$

(d) $\left(6y^4\right)\left(-2y^3\right)$

Use the associative and commutative properties as necessary to multiply numbers and multiply the variables.

$$\left(6y^4\right)\left(-2y^3\right) = 6(-2)y^4 y^3$$
$$= -12y^{4+3}$$
$$= -12y^7$$

(e) $\left(3p^5 q\right)\left(5p^6 q^3\right) = 3(5)p^5 p^6 q q^3 = 15p^{11}q^4$

(f) $x^3 y^2$

Because the bases are not the same, the product rule does not apply.

Define Zero as an Exponent: So far we have discussed only positive exponents. Let us consider how we might define a zero exponent. Suppose we multiply 5^3 by 5^0. By the product rule,

$$5^3 \cdot 5^0 = 5^{3+0} = 5^3.$$

For the product rule to hold true, 5^0 must equal 1, and so a^0 is defined as equal to 1 for any nonzero real number.

Definition: Zero Exponent

If a is any nonzero real number then,

$$a^0 = 1.$$

The symbol 0^0 is undefined.

EXAMPLE 2

(a) $11^0 = 1$

(b) $(-8)^0 = 1$

(c) $-8^0 = -\left(8^0\right) = -1$

(d) $5^0 + 11^0 = 1 + 1 = 2$

(e) $(9k)^0 = 1 \quad (k \neq 0)$

Self-Check 1

Simplify. Assume all variables represent nonzero real numbers.

1. $a^9 \cdot a^{12}$

2. $\left(-2x^2y\right)\left(-8x^3y^9\right)$

3. $3x^0$

4. $-2p^0 + 8z^0$

How should we define a negative exponent? Using the product rule again,

$$6^3 \cdot 6^{-3} = 6^{3+(-3)} = 6^0 = 1.$$

This indicates that 6^{-3} is the reciprocal of 6^3. But $\dfrac{1}{6^3}$ is the reciprocal of 6^3, and a number can have only one reciprocal. Thus, we must define 6^{-3} to equal $\dfrac{1}{6^3}$, so negative exponents are defined as follows.

Definition: Negative Exponent

For any natural number n and any nonzero real number a,

$$a^{-n} = \frac{1}{a^n}.$$

A negative exponent does not imply a negative number; negative exponents lead to reciprocals, for example:

$$4^{-2} = \frac{1}{4^2} = \frac{1}{16}.$$

EXAMPLE 3

In part (a)–(f), write the expression with only positive exponents. In part (g) simplify the expression.

(a) $3^{-2} = \dfrac{1}{3^2} = \dfrac{1}{9}$

(b) $6^{-1} = \dfrac{1}{6^1} = \dfrac{1}{6}$

(c) $(2p)^{-3} = \dfrac{1}{(2p)^3}, \quad p \neq 0$

(d) $2p^{-3} = 2\left(\dfrac{1}{p^3}\right) = \dfrac{2}{p^3}, \quad p \neq 0$

(e) $-m^{-4} = -\dfrac{1}{m^4}, \quad m \neq 0$

(f) $(-m)^{-4} = \dfrac{1}{(-m)^4}, \quad m \neq 0$

(g) $2^{-1} + 3^{-1}$

Since $2^{-1} = \dfrac{1}{2}$ and $3^{-1} = \dfrac{1}{3}$, $2^{-1} + 3^{-1} = \dfrac{1}{2} + \dfrac{1}{3} = \dfrac{3}{6} + \dfrac{2}{6} = \dfrac{5}{6}$.

Self-Check 2

Simplify. Assume all variables represent nonzero real numbers.

1. $(-2)^{-4}$ **2.** -2^4 **3.** $(3x)^{-2}$ **4.** $3^{-1} + 4^{-1}$

The Quotient Rule for Exponents: A quotient, such as $\dfrac{a^5}{a^3}$, can be simplified in much the same way as a product. (In all quotients of this type, assume that the denominator is not zero.) Using the definition of an exponent,

$$\frac{a^5}{a^3} = \frac{a \cdot a \cdot a \cdot a \cdot a}{a \cdot a \cdot a} = a \cdot a = a^2.$$

Notice that $5 - 3 = 2$. In the same way,

$$\frac{a^3}{a^5} = \frac{a \cdot a \cdot a}{a \cdot a \cdot a \cdot a \cdot a} = \frac{1}{a \cdot a} = \frac{1}{a^2}.$$

Here again, $5 - 3 = 2$. These examples suggest the following quotient rule for exponents.

Rule: Quotient Rule for Exponents

If a is any nonzero real number and m and n are integers, then

$$\frac{a^m}{a^n} = a^{m-n}.$$

EXAMPLE 4

Apply the quotient rule for exponents, and write the result using only positive exponents.

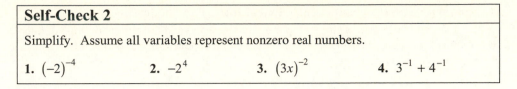

Numerator Exponent

Denominator Exponent

(a) $\dfrac{5^8}{5^2} = 5^{8-2} = 5^6$
Minus

(b) $\dfrac{a^7}{a^3} = a^{7-3} = a^4$

(c) $\dfrac{k^6}{k^{11}} = k^{6-11} = k^{-5} = \dfrac{1}{k^5}, \ k \neq 0$

(d) $\dfrac{3^{10}}{3^{-3}} = 3^{10-(-3)} = 3^{13}$

(e) $\dfrac{7^{-5}}{7^4} = 7^{-5-4} = 7^{-9} = \dfrac{1}{7^9}$

(f) $\dfrac{9^{-1}}{9^{-1}} = \dfrac{9^{-1}}{9^{-1}} = 9^{-1-(-1)} = 9^0 = 1$

(g) $\dfrac{z^{-4}}{z^{-12}} = z^{-4-(-12)} = z^8, \ (z \neq 0)$

(h) $\dfrac{x^5}{y^{15}}$

The quotient rule does not apply because the bases are different.

We can summarize the product and quotient rules: To multiply expressions such as a^m and a^n where the base is the same, keep the same base and add the exponents. To divide such expressions, keep the same base and subtract the exponent of the denominator from the exponent of the numerator.

Self-Check 3

Simplify and write the result using only positive exponents. Assume all variables represent nonzero real numbers.

1. $\dfrac{4^8}{4^2}$ **2.** $\dfrac{7^3}{7^{-5}}$ **3.** $\dfrac{s^2}{s^9}$ **4.** $\dfrac{r^{-8}}{r^{-3}}$

The Power Rules for Exponents: The expression $\left(5^3\right)^2$ can be simplified as $5^3 \cdot 5^3 = 5^{3+3} = 5^6$, where $6 = 3 + 3 = 2 \cdot 3$. This example suggests the first of the **power rules for exponents**; the other two parts can be demonstrated with similar examples.

Rule: Power Rule for Exponents

If a and b are real numbers and m and n are integers, then

$$\left(a^m\right)^n = a^{mn}, \qquad (ab)^m = a^m b^m, \quad \text{and} \quad \left(\frac{a}{b}\right)^m = \frac{a^m}{b^m} \quad (b \neq 0).$$

Here again we assume that zero never appears to a negative power.

EXAMPLE 5
Simplify. Assume all variables represent nonzero real numbers.

(a) $\left(p^4\right)^5 = p^{4 \cdot 5} = p^{20}$

(b) $\left(\dfrac{3}{4}\right)^3 = \dfrac{3^3}{4^3} = \dfrac{27}{64}$

(c) $(2z)^3 = 2^3 z^3 = 8z^3$

(d) $\left(5p^8\right)^2 = 5^2 p^{8 \cdot 2} = 25 p^{16}$

(e) $\left(\dfrac{-2c^6}{x^2}\right)^3 = \dfrac{(-2)^3 c^{6 \cdot 3}}{x^{2 \cdot 3}} = \dfrac{-8c^{18}}{x^6}, \quad (x \neq 0)$

Special Rules for Negative Exponents: Any nonzero number raised to the negative nth power is equal to the reciprocal of that number raised to the nth power. That is if $a \neq 0$ and $b \neq 0$,

$$a^{-n} = \left(\frac{1}{a}\right)^n \text{ and } \left(\frac{a}{b}\right)^{-n} = \left(\frac{b}{a}\right)^n.$$

EXAMPLE 6
Write the following expressions with only positive exponents and then evaluate.

(a) $5^{-3} = \left(\dfrac{1}{5}\right)^3 = \dfrac{1}{125}$

(b) $\left(\dfrac{1}{3}\right)^{-2} = 3^2 = 9$

(c) $\left(\dfrac{4}{5}\right)^{-2} = \left(\dfrac{5}{4}\right)^2 = \dfrac{25}{16}$

(d) $\left(\dfrac{2}{3}\right)^{-3} = \left(\dfrac{3}{2}\right)^3 = \dfrac{27}{8}$

Definitions/Rules: Definitions and Rules for Exponents

For all integers m and n and real numbers a and b:

Product Rule $a^m \cdot a^n = a^{m+n}$

Quotient Rule $\dfrac{a^m}{a^n} = a^{m-n} \quad (a \neq 0)$

Zero Exponent $a^0 = 1 \quad (a \neq 0)$

Negative Exponent $a^{-n} = \dfrac{1}{a^n} \quad (a \neq 0)$

Power Rules $\left(a^m\right)^n = a^{mn}, \; (ab)^m = a^m b^m, \; \text{and} \; \left(\dfrac{a}{b}\right)^m = \dfrac{a^m}{b^m} \quad (b \neq 0)$

Special Rules $a^{-n} = \left(\dfrac{1}{a}\right)^n \quad (a \neq 0) \; \text{and} \; \left(\dfrac{a}{b}\right)^{-m} = \dfrac{b^m}{a^m} \quad (a, b \neq 0)$

Self-Check 4

Simplify. Assume all variables represent nonzero real numbers.

1. $\left(3x^2 y^5\right)^2$ **2.** $\left(\dfrac{2t^4}{u^7}\right)^4$ **3.** x^{-4} **4.** $\left(\dfrac{2}{3}\right)^{-3}$

Simplify Exponential Expressions: With the rules of exponents developed so far, we can simplify expressions that involve one or more rules.

EXAMPLE 7

Simplify each expression so that no negative exponents appear in the final result. Assume that all variables represent nonzero real numbers.

(a) $2^4 \cdot 2^{-6} = 2^{4+(-6)} = 2^{-2} = \dfrac{1}{2^2} = \dfrac{1}{4}$

(b) $y^3 \cdot y^{-8} \cdot y^{-2} = y^{3+(-8)+(-2)} = y^{-7} = \dfrac{1}{y^7}$

(c) $\left(z^{-4}\right)^{-3} = z^{-4(-3)} = z^{12}$

(d) $\left(p^8\right)^{-2} = p^{8 \cdot (-2)} = p^{-16} = \dfrac{1}{p^{16}}$

(e) $\dfrac{x^{-5}y^6}{x^3 y^{-4}} = x^{-5-3}y^{6-(-4)} = x^{-8}y^{10} = \dfrac{y^{10}}{x^8}$

Self-Check 5

Simplify. Assume all variables represent nonzero real numbers.

1. $z^{-4} \cdot z^{-11} \cdot z^5$ **2.** $\left(2^{-3}\right)^2$ **3.** $x^{-4} \cdot x^6 \cdot x^{-2}$ **4.** $\dfrac{p^{-11}q^3}{p^{-5}q^4}$

Scientific Notation: Scientists often need to use numbers that are very large or very small. It is more convenient to write these numbers using the rules for exponents.

In **scientific notation**, a number is written with the decimal point after the first nonzero digit and multiplied by a power of 10, as indicated in the following definition.

Definitions: Scientific Notation

A number is written in **scientific notation** when it is expressed in the form

$$a \times 10^n$$

where $1 \le |a| < 10$, and n is an integer.

For example, in scientific notation,

$$5000 = 5 \times 1000 = 5 \times 10^3.$$

The following numbers are *not* in scientific notation.

0.345×10^5	14.1×10^{-8}
0.345 is less than 1.	14.1 is greater than 10.

To write a number in scientific notation, use the steps given below. (If the number is negative, ignore the minus sign, go through the steps, and attach the minus sign to the result.)

Procedure: Converting to Scientific Notation

Step 1   **Position the decimal point.** Place a caret, ^, to the right of the first nonzero digit, where the decimal point will be placed.

Step 2   **Determine the numeral for the exponent.** Count the number of digits from the caret to the decimal point. This number gives the absolute value of the exponent on ten.

Step 3   **Determine the sign for the exponent.** Decide whether multiplying by 10^n should make the result of Step 1 larger or smaller. The exponent should be positive to make the result larger; it should be negative to make the result smaller.

EXAMPLE 8
Write each number in scientific notation.

(a) 930,000

Place a caret to the right of the 9 (the first nonzero digit) to mark the new location of the decimal point.

$$9_\wedge 30{,}000$$

Count from the caret to the decimal point, which is understood to be after the last digit.

0.9 30,000. ←Decimal point

Count 5 places

Since the number 9.3 is to be made larger, the exponent on 10 is positive.

$$930{,}000 = 9.3 \times 10^5$$

(b) 0.0000024

Count from right to left.

$$0.000002\,4$$

6 places

Since the number 2.4 is to be made smaller, the exponent on 10 is negative.

$$0.0000024 = 2.4 \times 10^{-6}$$

To convert a number written in scientific notation to standard notation, just work in reverse.

Procedure: Converting from Scientific Notation

Multiplying a number by a positive power of 10 makes the number larger, so move the decimal point to the right if n is positive in 10^n.

Multiplying by a negative power of 10 makes a number smaller, so move the decimal point to the left if n is negative.

If n is zero, leave the decimal point where it is.

EXAMPLE 9

Write the following numbers without scientific notation.

(a) 8.21×10^6

$$8.210000$$

6 places

The decimal point was moved 6 placed to the right. (It was necessary to attach 4 zeros.)

$$8.21 \times 10^6 = 8,210,000$$

(b) 5.7×10^{-5}

$$00005.7$$

5 places

The decimal point was moved 5 placed to the left. Therefore, $5.7 \times 10^{-5} = 0.000057$.

(c) $2.589 \times 10^0 = 2.589$

Scientific notation can be used to solve problems. First write the numbers in scientific notation and then perform the calculations using the rules for exponents.

EXAMPLE 10

The Greek astronomer Aristarchis calculated the distance from the Earth to the Sun to be about 19 times that of the distance between the Earth and the Moon. If the distance from the Earth to the Moon is approximately a quarter million miles then

(a) what is this distance in scientific notation?

$$250,000,000 = 2.5 \times 10^8 \text{ miles}$$

(b) what did Aristarchus approximate the distance from the Earth to the Sun to be in scientific notation?

$$\left(2.5 \times 10^8\right) \cdot 19 = (2.5 \cdot 19) \times 10^8 = 47.5 \times 10^8 = \left(4.75 \times 10^1\right) \times 10^8$$

$$= 4.75 \times \left(10^1 \cdot 10^8\right) = 4.75 \times 10^{1+8} = 4.75 \times 10^9 \text{ miles}$$

Self-Check 6

Convert the following numbers into scientific notation.

1. −23,000,000

2. 0.00000056

Perform the indicated operation by using rules for exponents and properties of real numbers. Convert your answer from scientific notation.

3. $\left(3 \times 10^{-4}\right) \cdot \left(2 \times 10^6\right)$

4. $\dfrac{8 \times 10^{-1}}{2 \times 10^3}$

Self-Check Answers

1.1 a^{21}	**1.2** $16x^5 y^{10}$	**1.3** 3	**1.4** 6
2.1 $\dfrac{1}{16}$	**2.2** $-\dfrac{1}{16}$	**2.3** $\dfrac{1}{9x^2}$	**2.4** $\dfrac{7}{12}$
3.1 4^6	**3.2** 7^8	**3.3** $\dfrac{1}{s^7}$	**3.4** $\dfrac{1}{r^5}$
4.1 $9x^4 y^{10}$	**4.2** $\dfrac{16t^{16}}{u^{28}}$	**4.3** $\dfrac{1}{x^4}$	**4.4** $\dfrac{27}{8}$
5.1 $\dfrac{1}{z^{10}}$	**5.2** $\dfrac{1}{64}$	**5.3** 1	**5.4** $\dfrac{1}{p^6 q}$
6.1 -2.3×10^7	**6.2** 5.6×10^{-7}	**6.3** 600	**6.4** 0.0004

3.1 EXERCISES

Decide whether the expression has been simplified correctly in exercises 1-4. If not, correct it.

1. $(ab)^3 = ab^3$

2. $(-2x)^5 = (-2)^5 x^5$

3. $x^2 x^5 = x^{10}$

4. $\dfrac{y^{12}}{y^3} = y^4 \quad (y \neq 0)$

5. State the product and quotient rules for exponents in your own words. Give examples with your explanation.

6. State the three power rules for exponents in your own words. Give examples with your explanation.

For exercises 7-26, use the product and/or quotient rules as needed to simplify each expression. Write the answer with only positive exponents. Assume that all variables represent nonzero real numbers.

7. $x^5 \cdot x^9$

8. $y^2 \cdot y^{10}$

9. $x^3 \cdot x^{10} \cdot x^{-4}$

10. $b^8 \cdot b^{-7} \cdot b^9$

11. $\left(3x^2 y^6\right)\left(-8x^3 y^7\right)$

12. $\left(-6xy^9\right)\left(-10x^5 y^9\right)$

13. $\dfrac{q^{12}}{q^5}$

14. $\dfrac{p^9}{p^4}$

15. $\dfrac{z^{-8}}{z^{-12}}$

16. $\dfrac{g^{-4}}{g^{-8}}$

17. $\dfrac{r^{12}r^{-5}r^{-4}}{r^{-3}r^{-6}r^{0}}$

18. $\dfrac{z^{-5}z^{-5}z^{0}}{z^{2}z^{-6}z^{-4}}$

19. $9k^{3}\left(-3k\right)\left(5k^{-6}\right)^{0}$

20. $2a^{3}\left(-5a\right)\left(9a^{-9}\right)^{0}$

21. $-7\left(4x^{4}\right)\left(5x\right)$

22. $10\left(6z^{5}\right)\left(2xz^{3}\right)$

23. $\dfrac{\left(4xy\right)y^{3}}{12x^{3}y^{6}}$

24. $\dfrac{\left(-16ab\right)b^{9}}{8a^{5}b^{10}}$

25. $\dfrac{12x^{-6}y^{-3}}{\left(4x^{-4}y^{-4}\right)\left(3x^{-2}y^{-3}\right)}$

26. $\dfrac{-16a^{3}b^{-8}}{\left(8ab^{-6}\right)\left(-2a^{2}b\right)}$

27. Your friend evaluated $3^{2} \cdot 3^{6}$ as 9^{8}. Explain to your friend why this is not correct.

28. Consider the expression $-a^{n}$ and $\left(-a\right)^{n}$. In some cases they are equal and in some cases they are not. Using $n = 2,\ 3,\ 4,\ 5$ and 6 and $a = 2$, draw conclusions as to when they are equal and when they are opposites.

Evaluate exercises 29-49.

29. $\left(\dfrac{3}{4}\right)^{2}$

30. $\left(\dfrac{2}{3}\right)^{3}$

31. 5^{-3}

32. 2^{-2}

33. -5^{-2}

34. $\left(-3\right)^{-2}$

35. $\left(-4\right)^{-3}$

36. $\dfrac{1}{4^{-2}}$

37. $\dfrac{1}{7^{-1}}$

38. $\dfrac{-2^{-2}}{5^{-1}}$

39. $\dfrac{5^{-1}}{-4^{-3}}$

40. $\left(\dfrac{3}{4}\right)^{-2}$

41. $\left(\dfrac{5}{2}\right)^{-3}$

42. $6^{-1} + 4^{-1}$

43. $2^{-1} - 3^{-1}$

44. $\left(7 - 5\right)^{-1}$

45. $\left(10 - 14\right)^{-1}$

46. $\dfrac{5^{-5}}{5^{-3}}$

47. $\dfrac{3^{-8}}{3^{-5}}$

48. $\dfrac{10^{-1}}{-10}$

49. $\dfrac{7^{-1}}{-7}$

Evaluate each expression in exercises 50-75. Assume that all variables represent nonzero real numbers.

50. $\left(2^{-4} \cdot 3^{-1}\right)^{2}$

51. $\left(2^{-3} \cdot 3^{-2}\right)^{2}$

52. $\left(3^{-2} \cdot 5^{-3}\right)^{-1}$

53. $\left(2^{-6} \cdot 3^{-3}\right)^{-1}$

54. $\left(x^{2}\right)^{-4}x^{5}$

55. $\left(a^{3}\right)^{-5}a^{7}$

56. $-3t^{-4}\left(t^3\right)^7$

57. $-4x^{-1}\left(x^5\right)^3$

58. $\left(3d^{-1}\right)^3\left(d^3\right)^{-5}$

59. $\left(5y^{-1}\right)^2\left(y^4\right)^{-1}$

60. $\left(x^{-3}y^5\right)^{-1}$

61. $\left(a^{-5}b^6\right)^{-2}$

62. $\dfrac{\left(k^{-4}\right)^0}{7k^{-5}}$

63. $\dfrac{\left(z^{-2}\right)^0}{8z^{-7}}$

64. $\dfrac{5b^3\left(b^{-1}\right)^4}{\left(b^{-3}\right)^{-2}}$

65. $\dfrac{8p^{-3}\left(p^{-5}\right)^{-4}}{2p^6}$

66. $\dfrac{\left(-z^{-3}\right)^2}{2\left(z^{-8}\right)^{-1}}$

67. $\dfrac{7\left(-y^{-1}\right)^{-5}}{2\left(y^{-4}\right)^3}$

68. $\dfrac{\left(3x\right)^2 y^{-4}}{\left(xy\right)^{-2}}$

69. $\dfrac{\left(4ab\right)^{-2}}{2^3 a^5 b^{-8}}$

70. $\dfrac{\left(3x\right)^3 x^4}{x^{-1}x^{-6}}\left(2x^{-3}\right)^{-2}$

71. $\dfrac{\left(4y\right)^2 y^{-7}}{y^{-4}y^8}\left(3y^{-4}\right)^3$

72. $\left(\dfrac{2x^{-4}}{x^5}\right)^{-1}\cdot\dfrac{3}{x}$

73. $\left(\dfrac{3a^{-8}}{a^{-4}}\right)^{-2}\cdot\dfrac{a^5}{2}$

74. $\left(\dfrac{2x}{x^3}\right)^4\cdot\left(\dfrac{3x^5}{x^{-7}}\right)^{-1}$

75. $\left(\dfrac{2y^4}{3x^5}\right)^{-4}\cdot\left(\dfrac{3x^{-1}}{6y^{-3}}\right)^{-4}$

Write each number in scientific notation in exercises 76-83.

76. 640

77. 5300

78. 0.670

79. 0.000345

80. 0.0000078

81. 0.127

82. −48,200

83. −0.00042

Write each of the following in standard notation in exercises 84-91.

84. 3.4×10^6

85. 3.21×10^2

86. 6.25×10^{-3}

87. 2.46×10^{-5}

88. -2×10^6

89. -3×10^4

90. 1.3×10^{-5}

91. 4.1×10^{-7}

Use the rules for exponents to find each value in exercises 92-99.

92. $\dfrac{15 \times 10^5}{3 \times 10^9}$

93. $\dfrac{35 \times 10^8}{7 \times 10^{14}}$

94. $\dfrac{2 \times 10^{-4}}{8 \times 10^5}$

95. $\dfrac{4 \times 10^{-9}}{20 \times 10^7}$

96. $\dfrac{0.06 \times 2400}{0.0002}$

97. $\dfrac{0.008 \times 20,000}{0.00016}$

98. $\dfrac{15,000 \times 0.016}{200 \times 0.04}$

99. $\dfrac{420,000 \times 0.00006}{0.00021 \times 20,000}$

▦*Solve each problem in exercises 100–103. Use a calculator as necessary.*

100. A parsec, a unit of length used in astronomy, is 19×10^{12} miles. The mean distance of Uranus from the sun is 1.8×10^7 miles. How many parsecs is Uranus from the sun?

101. An inch is approximately 1.57828×10^{-5} mile. Find the reciprocal of this number to determine the number of inches in a mile.

102. The speed of light is approximately 3×10^{10} centimeters per second. How long will it take light to travel 9×10^{12} centimeters?

103. The average distance from Earth to the sun is 9.3×10^7 miles. How long would it take a rocket, traveling at 2.9×10^3 miles per hour, to reach the sun?

3.2 | Addition and Subtraction of Polynomials

The Basic Definitions for Polynomials: Just as whole numbers are the basis of arithmetic, *polynomials* are fundamental in algebra. To understand polynomials, we first review several words from Chapter 1: A *term* is a number, a variable, or the product or quotient of a number and one or more variables raised to nonnegative powers.

Examples of terms include:

$$8x^3, \quad -5p^5, \quad k, \quad -r, \quad \text{and} \quad 3x^2 y.$$

The number in the product is called the *numerical coefficient,* or just the *coefficient.* In the term $8x^3$, the coefficient is 8. In the term $-5p^5$, it is -5. The coefficient of the term k is understood to be 1. The coefficient of $-r$ is -1. More generally, any factor in a term is the coefficient of the product of the remaining factors. For example, $3x^2$ is the coefficient of y in the term $3x^2 y$, and $3y$ is the coefficient of x^2 in $3x^2 y$.

Any combination of variables or constants (numerical values) joined by the basic operations of addition, subtraction, multiplication, and division (except by 0), or taking roots is called an **algebraic expression**. The simplest kind of algebraic expression is a **polynomial**.

Definition: Polynomial

A **polynomial** is a term or a finite sum of terms in which all variables have whole number exponents and no variables appear in denominators.

Examples of polynomials include:

$$2x - 5, \quad 4m^4 - 5mp^2 + 1, \quad \text{and} \quad -2x^2 y^3.$$

Even though the expression $2x - 5$ involves subtraction, it is called a sum of terms, since it could be written as $2x + (-5)$.

Some examples of expressions that are *not* polynomials are:

$$x^{-2} + 5x^{-3}, \quad \sqrt{5 - x}, \quad \text{and} \quad \frac{1}{x}.$$

A polynomial containing only the variable x is called a **polynomial in x**. A polynomial in one variable is written in **descending powers** of the variable if the exponents on the terms of the polynomial decrease from left to right. For example,

$$x^5 - 5x^2 + 17x - 4$$

is a polynomial in descending powers of x. The term -4 in the polynomial above can be thought of as $-4x^0$, since $-4x^0 = -4(1) = -4$.

EXAMPLE 1
Write each of the following in descending powers of the variable.

(a) $-y - 7y^2 + 8y^6 - 10y^4 + 16$ would be written as $8y^6 - 10y^4 - 7y^2 - y + 16$.

(b) $-8 - x^2 + 8x^3 - 10x^4$ would be written as $-10x^4 + 8x^3 - x^2 - 8$.

Polynomials with a specific number of terms are so common that they are given special names. A polynomial of exactly three terms is a **trinomial**, and a polynomial with exactly two terms is a **binomial**. A single-term polynomial is a **monomial**.

The Degree of a Polynomial: The degree of a term with one variable is the exponent on the variable. For example, the degree of $8x^3$ is 3, the degree of $-x^5$ is 5, and the degree of $14x$ is 1. The degree of a term in more than one variable is defined to be the *sum of the exponents* of the variables. For example, the degree of $4x^4y^6$ is 10, because $4 + 6 = 10$.

The greatest degree of any of the terms in a polynomial is called the **degree of the polynomial**. For example, the degree of $5x^4 + 5x^3 - 10x^2 + 1$ is 4, because the greatest degree of any term is 4 (the degree of $5x^4$).

EXAMPLE 2
Find the degree of each polynomial.

    Greatest exponent is 2.
    ↓

(a) $5x^2 - 2x + 1$
    The greatest exponent is 2, so the polynomial is of degree 2.

(b) $19x^4 + 12x^6 - 6x^2$ (c) $7x$
    This polynomial is of degree 6.   The degree is 1, since $7x = 7x^1$.

(d) -9
    Since $-9 = -9x^0$, the degree is 0. Any constant term, other than zero, has degree 0. The number 0 has no degree, since 0 times a variable to any power is 0.

(e) $6a^3b^7$
    The degree is the sum of the exponents, $3 + 7 = 10$.

Self-Check 1

Identify the degree of the following polynomials and rewrite (if necessary) in descending powers of the variable.

1. $3a^2 - 4a^6 + 2$ **2.** $4x^5 - 3x^2 + 2x - 9$

Determine if the following are polynomials. If it is a polynomial, give its special name. If not, state why it is not a polynomial.

3. $7x^3 + 1$ **4.** $7x^{-3} + 1$

Add and Subtract Polynomials: We use the distributive property to simplify polynomials by combining terms. For example, simplify $x^3 + 5x^2 + 3x^2 - 8$ as follows.

$$x^3 + 5x^2 + 3x^2 - 8 = x^3 + (5 + 3)x^2 - 8 \qquad \text{Distributive property}$$
$$= x^3 + 8x^2 - 8$$

On the other hand, the terms in the polynomial $5x + 3x^2$ cannot be combined. As these examples suggest, only terms containing exactly the same variables to the same powers may be combined. As mentioned in Chapter 1, such terms are called *like terms*.

EXAMPLE 3
Combine terms.

(a) $-8x^4 + 12x^4 - x^4 = (-8 + 12 - 1)x^4 = 3x^4$

(b) $3x + 7y - 10x + 14y$

Use the associative and commutative properties to rewrite the expression with all the x-terms together and all the y-terms together.

$$3x + 7y - 10x + 14y = 3x - 10x + 7y + 14y$$

Now combine like terms.

$$= -7x + 21y$$

Since $-7x$ and $21y$ are unlike terms, no further simplification is possible.

Rule: Adding Polynomials

To add two polynomials, combine like terms

Polynomials can be added horizontally or vertically.

EXAMPLE 4
Add: $\left(2x^3 - 3x + 2\right) + \left(-4x^2 + 12x - 5\right)$

$$\left(2x^3 - 3x + 2\right) + \left(-4x^2 + 12x - 5\right) = 2x^3 - 4x^2 - 3x + 12x + 2 - 5$$
$$= 2x^3 - 4x^2 + 9x - 3$$

Add these same two polynomials vertically by placing like terms in columns.

$$
\begin{array}{r}
2x^3 \qquad\ - 3x + 2 \\
-4x^2 + 12x - 5 \\
\hline
2x^3 - 4x^2 + \ 9x - 3
\end{array}
$$

In Chapter 1, subtraction of real numbers was defined as

$$a - b = a + (-b).$$

That is, add the first number and the negative (or opposite) of the second. We can give a similar definition for subtraction of polynomials by defining the **negative of a polynomial** as that polynomial with every sign changed.

Rule: Subtracting Polynomials

To subtract two polynomials, add the first polynomial and the negative of the *second* polynomial.

EXAMPLE 5

(a) Subtract: $\left(-7y^2 + 4y - 8\right) - \left(-8y^2 + 2y + 6\right)$.

Change every sign in the second polynomial and add.

$$\left(-7y^2 + 4y - 8\right) - \left(-8y^2 + 2y + 6\right) = -7y^2 + 4y - 8 + 8y^2 - 2y - 6$$

Now add by combining like terms.

$$= -7y^2 + 8y^2 + 4y - 2y - 8 - 6$$
$$= y^2 + 2y - 14$$

Check by adding the sum, $y^2 + 2y - 14$, to the second polynomial. The result should be the first polynomial.

(b) Subtract the two polynomials in part (a) vertically.
Write the first polynomial above the second, lining up like terms in columns.

$$-7y^2 + 4y - 8$$
$$\underline{-8y^2 + 2y + 6}$$

Change all the signs in the second polynomial, and add.

$$-7y^2 + 4y - 8$$
$$\underline{8y^2 - 2y - 6} \qquad \text{All signs changed.}$$
$$y^2 + 2y - 14 \qquad \text{Add in columns.}$$

Self-Check 2

Add or subtract the following polynomials as indicated.

1. $\left(3x^2 - 10x + 2\right) + \left(-8x^2 + 10x + 1\right)$ 2. $\left(4y^3 + 3y^2 - 8\right) + \left(2y^2 - 7y + 8\right)$

3. $\left(9z^2 - 8z + 12\right) - \left(8z^2 + 6z + 17\right)$ 4. $\left(12 + 6y^2 + 4y\right) - \left(9y^2 + 12y - 8\right)$

Self-Check Answers

1.1 Sixth degree; $-4a^6 + 3a^2 + 2$

1.2 Fifth degree; already in descending order

1.3 Polynomial; binomial

1.4 Not a polynomial; Exponents of variables are not all whole numbers.

2.1 $-5x^2 + 3$ **2.2** $4y^3 + 5y^2 - 7y$ **2.3** $z^2 - 14z - 5$ **2.4** $-3y^2 - 8y + 20$

3.2 EXERCISES

Write each polynomial in descending powers of the variable in exercises 1-6.

1. $4x^3 + 3x - 2x^2 + 12$ 2. $7z^4 + 3z^3 - 21z^5 + 10z$

3. $-8p^3 - 3p^2 + 19p$ 4. $-4q^3 + 3q - 2q^4 + 1$

5. $-y^5 + 3y - 2y^2 + 10$ 6. $4 + 3x - 2x^2$

For exercises 7-14, identify each polynomial as a monomial, binomial, trinomial, or none of these. Also give the degree.

7. 16   **8.** -8 **9.** $9c - 2$

10. $b^8 + 3b^6$ **11.** $7x^3 + 4x^2 + 3$ **12.** $2z^2 - 4z + 1$

13. $4x^4y^3 - 9x^3y^3 - 21x^2y + xy$ **14.** $-6a^5b^2 - 8a^4b^2 + 5a^2b + ab$

15. Which one of the following is a trinomial in descending powers, having degree 5?
     (a) $5x^5 + 4x^4 + 3x^3 + 2x^2 + 1$ (b) $3x^3 - 2x^5 + 7$
     (c) $6 + 7x^3 + 4x^5$ (d) $6x^5 - 5x^4 + 3$

16. Give an example of a polynomial of five terms in the variable x, having degree 6, written in descending powers, lacking a fourth degree term.

Give the numerical coefficient and the degree of each term in exercises 17-24.

17. $4p$ **18.** $-6q$ **19.** $-12p^3$ **20.** $-34x^9$

21. x^5 **22.** y^{10} **23.** $-ab^2$ **24.** $-x^2y$

Combine terms in exercises 25-45.

25. $5p^4 + 6p^4$ **26.** $9x^3 - 2x^3$ **27.** $-r^3 + 6r^3 + 7r^3$

28. $4x^4 - 5x^4 + 6x^4$ **29.** $y + y + y + y + y + y$ **30.** $x - x - x + x$

31. $2y^3 - y + 7y^3$ **32.** $3x^2 - 2 + 10 - x^2$ **33.** $3x + 5x^2 + 6x^2 - 2$

34. $5x^2 + 6x - 7x^2 - 2$ **35.** $k^4 - 3k^3 + k^2 - 7k^4 + k^3$

36. $5y^3 + 5y^2 - 5y - y^2 + 9y^2$ **37.** $\left[6 - (8 + 7p)\right] + (9p + 10)$

38. $\left[5x - (2x + 8)\right] + (7x - 9)$ **39.** $\left[(2 + 7y) - (4y + 3)\right] - (2y + 11)$

40. $\left[(8b - 7) - (-2 + b)\right] - (5b + 6)$ **41.** $\left(5p^2 + 6p - 8\right) + \left(8p^2 - 9p^3 + 2p\right)$

42. $\left(y^3 + 4y + 1\right) + \left(7y^3 - 6y^2 + 2y - 8\right)$ **43.** $\left(5x^5 - 3x^4 + x^3 - 4\right) + \left(x^4 - 5x^3 + 8\right)$

44. $\left(u^2 + 5u\right) + \left[2u - \left(7u^2 + 6u + 9\right)\right]$ **45.** $\left(6a - 2a\right) - \left[8a - (6a + 7)\right]$

46. Define *polynomial* in your own words. Give examples. Include the words *term*, *monomial*, *binomial*, and *trinomial* in your explanation.

47. Write a paragraph explaining how to add and subtract polynomials. Give examples.

48. Consider the exponent in the expression 4^3. Explain why the degree of 4^3 is not 3. What is its degree?

Add or subtract as indicated in exercises 49-56.

49. Add.

$$27y - 9$$
$$\underline{-2y + 6}$$

50. Add.

$$14n - 8$$
$$\underline{3n + 16}$$

51. Add.

$$-13p^2 + 5p - 6$$
$$\underline{-3p^2 + 6p + 9}$$

52. Add.

$$-9b^2 + 7b - 12$$
$$\underline{4b^2 + 6b + 2}$$

53. Subtract.

$$7x^2 - 12x - 10$$
$$\underline{-9x^2 + 3x - 6}$$

54. Subtract.

$$-5z^2 + 4z - 3$$
$$\underline{3z^2 - 7z + 2}$$

55. Subtract.

$$6q^2 - 3q + 4$$
$$\underline{-8q^2 + 2q - 8}$$

56. Subtract.

$$5y^3 - 8y^2 + 6y$$
$$\underline{2y^3 + 2y^2 \quad + 9}$$

Perform the operations in exercises 57-64.

57. Subtract $6x^2 - 2x + 4$ from $8x^2 - 5x + 8$.

58. Subtract $-\left(3x + 6y^2 + 6z\right)$ from $\left[\left(6z - 9x + y^2\right) + \left(y^2 - 3z\right)\right]$.

59. $\left(-9m^2 + 6n^2 - 9n\right) - \left[\left(2m^2 - 7n^2 + 7n\right) + \left(-7m^2\right) + 6n^2\right]$

60. $\left[-\left(5x^2 - 9x + 8x^3\right) - \left(6x^2 + 8x + 6x^3\right)\right] + x^2$

61. $\left[-\left(y^4 - 7y^2 + 3\right) - \left(y^4 + 2y^2 + 3\right)\right] + \left(5y^4 - 5y^2 - 8\right)$

62. $\left(3p - [9p - 7]\right) - \left[\left(6p - (9 - 2p)\right) + 13p\right]$

63. $-\left(6z^2 + 9z - \left[3z^2 - 5z\right]\right) + \left[\left(5z^2 - \left[2z - z^2\right] + 6z^2\right)\right]$

64. $9k - \left(9k - \left[6k - (3k - 9k)\right] + 11k - (12k - 15k)\right)$

3.3 **Multiplication of Polynomials**

Multiply Terms: Recall that the product of the two terms $4x^5$ and $7x^6$ is found by using the commutative and associative properties, along with the rules for exponents.

$$\left(4x^5\right)\left(7x^6\right) = 4 \cdot 7 \cdot x^5 \cdot x^6$$
$$= 28x^{5+6}$$
$$= 28x^{11}$$

EXAMPLE 1
Find the following products.

(a) $\left(-3a^4\right)\left(5a^5\right) = (-3)(5)a^4 \cdot a^5 = -15a^9$

(b) $\left(4m^2n^4\right)\left(11m^5n^8\right) = (4)(11)m^2 \cdot m^5 \cdot n^4 \cdot n^8 = 44m^7n^{12}$

Multiply Any Two Polynomials: The distributive property can be used to extend this process to find the product of any two polynomials.

EXAMPLE 2

Find the following products.

(a) $-3\left(5x^4 - 10x^2\right) = -3\left(5x^4\right) - 3\left(-10x^2\right)$ Distributive property

$$= -15x^4 + 30x^2$$

(b) $2x^2\left(-3x^2 + 5x - 8\right) = 2x^2\left(-3x^2\right) + 2x^2\left(5x\right) - 2x^2\left(8\right)$

$$= -6x^4 + 10x^3 - 16x^2$$

(c) $\left(2x - 5\right)\left(5x^2 + x\right)$

Use the distributive property to multiply each term of $5x^2 + x$ by $2x - 5$.

$$\left(2x - 5\right)\left(5x^2 + x\right) = \left(2x - 5\right)\left(5x^2\right) + \left(2x - 5\right)\left(x\right)$$

Here $2x - 5$ has been treated as a single expression so that the distributive property could be used. Now use the distributive property two more times.

$$= 2x\left(5x^2\right) + \left(-5\right)\left(5x^2\right) + \left(2x\right)\left(x\right) + \left(-5\right)\left(x\right)$$

$$= 10x^3 - 25x^2 + 2x^2 - 5x$$

$$= 10x^3 - 23x^2 - 5x$$

(d) $3x^2\left(x + 2\right)\left(x - 4\right) = 3x^2\left[\left(x + 2\right)\left(x\right) + \left(x + 2\right)\left(-4\right)\right]$

$$= 3x^2\left[x^2 + 2x - 4x - 8\right]$$

$$= 3x^2\left[x^2 - 2x - 8\right]$$

$$= 3x^4 - 6x^3 - 24x^2$$

It is often easier to multiply polynomials by writing them vertically. To find the product from Example (c), $\left(2x - 5\right)\left(5x^2 + x\right)$, vertically, proceed as follows. (Notice how this process is similar to that of finding the product of two numbers, such as 25×32.)

1. Multiply x and $2x - 5$

$$2x\ -5$$
$$5x^2 +\ x$$
$$x\left(2x - 5\right) \rightarrow 2x^2 - 5x$$

2. Multiply $5x^2$ and $2x - 5$.

   Line up like terms of the

   products in columns.

$$2x\ -5$$
$$5x^2 +\ x$$
$$\overline{2x^2 -5x}$$
$$5x^2\left(2x - 5\right) \rightarrow \underline{10x^3 -\ 25x^2}$$

3. Combine like terms.

$$10x^3 - 23x^2 - 5x$$

EXAMPLE 3

Find the product, $(3a - 4b)(7a + b)$, vertically.

$$
\begin{array}{r}
3a - 4b \\
7a + b \\
\hline
\end{array}
$$

$$3ab - 4b^2 \quad \leftarrow b(3a - 4b)$$
$$\underline{21a^2 - 28ab} \qquad \leftarrow 7a(3a - 4b)$$
$$21a^2 - 25ab - 4b^2 \quad \text{Combine like terms.}$$

Self-Check 1

Find the following products.

1. $\left(-3x^2 y^3\right)\left(12x^3 y^6\right)$

2. $3x^2\left(4x^3 - 9x\right)$

3. $(4x - 9)\left(2x^2 + 3x\right)$

4. $(2a - b)(3a + b)$

Multiply Binomials: We can find the product of two binomials using the distributive property as follows.

$$(2x - 5)(4x + 7) = 2x(4x + 7) - 5(4x + 7)$$
$$= 2x(4x) + 2x(7) - 5(4x) - 5(7)$$
$$= 8x^2 + 14x - 20x - 35$$

Before combining like terms to find the simplest form of the answer, let us check the origin of each of the four terms in the sum. First, $8x^2$ is the product of the two *first* terms.

$$(\underline{2x} - 5)(\underline{4x} + 7) \qquad 2x(4x) = 8x^2 \qquad \text{First terms}$$

To get $14x$, the outside terms are multiplied.

$$(\underline{2x} - 5)(4x + \underline{7}) \qquad 2x(7) = 14x \qquad \text{Outside terms}$$

To get $-20x$, the inside terms are multiplied.

$$(2x - \underline{5})(\underline{4x} + 7) \qquad -5(4x) = -20x \qquad \text{Inside terms}$$

Finally, -35 comes from the last terms.

$$(2x - \underline{5})(4x + \underline{7}) \qquad -5(7) = -35 \qquad \text{Last terms}$$

The product is found by combining these four results.

$$(2x - 5)(4x + 7) = 8x^2 + 14x - 20x - 35$$
$$= 8x^2 - 6x - 35$$

To keep track of the order of multiplying these terms, we use the initials FOIL (First, Outside, Inside, Last). All the steps of the FOIL method can be done as follows. Try to do as many of these steps as possible in your head.

EXAMPLE 4

Use the FOIL method to multiply the binomials.

(a) $(3x + 2)(7x - 6) = 21x^2 - 18x + 14x - 12$

$\qquad\qquad\qquad\qquad$ First Outside Inside Last

$\qquad\qquad\qquad = 21x^2 - 4x - 12$

(b) $(4a - 5b)(6a + 5b) = 24a^2 + 20ab - 30ab - 25b^2$

$\qquad\qquad\qquad\qquad = 24a^2 - 10ab - 25b^2$

Self-Check 2

Use the FOIL method to multiply the binomials.

1. $(2x + 1)(3x + 5)$ $\qquad\qquad\qquad$ **2.** $(6y - 1)(9y + 7)$

3. $(6x + y)(3x - 7y)$ $\qquad\qquad\qquad$ **4.** $(3a - 4b)(7a - 5b)$

Multiply the Sum and Difference of Two Terms: Some types of binomial products occur frequently. By the FOIL method, the product of the sum and difference of the same two terms $(x + y)(x - y)$ is

$$(x + y)(x - y) = x^2 - xy + xy - y^2$$
$$= x^2 - y^2.$$

Rule: Product of the Sum and Difference of Two Terms

The **product of the sum and difference of two terms** is the difference of the squares of the terms, or

$$(x + y)(x - y) = x^2 - y^2.$$

EXAMPLE 5

Find the following products.

(a) $(x + 5)(x - 5) = x^2 - 5^2 = x^2 - 25$

(b) $(3x + 7)(3x - 7) = (3x)^2 - 7^2 = 3^2 x^2 - 49 = 9x^2 - 49$

(c) $(5a + 3b)(5a - 3b) = (5a)^2 - (3b)^2 = 5^2 a^2 - 3^2 b^2 = 25a^2 - 9b^2$

(d) $3x^2(x + 2)(x - 2) = 3x^2(x^2 - 4) = 3x^4 - 12x^2$

The special product $(x + y)(x - y) = x^2 - y^2$ can be used to perform some multiplication problems. For example,

$$101 \times 99 = (100 + 1)(100 - 1) = 100^2 - 1$$
$$= 10{,}000 - 1$$
$$= 9{,}999.$$

Once these patterns are recognized, multiplications like this can be done mentally.

The Square of a Binomial: Another special binomial product is the *square of a binomial*. To find the square of $x + y$, or $(x + y)^2$, multiply $x + y$ and $x + y$.

$$(x + y)^2 = (x + y)(x + y)$$
$$= x^2 + xy + xy + y^2$$
$$= x^2 + 2xy + y^2$$

A similar result is true for the square of a difference.

Rule: Square of a Binomial

The **square of a binomial** is the sum of the square of the first term, twice the product of the two terms, and the square of the last term.

$$(x + y)^2 = x^2 + 2xy + y^2$$
$$(x - y)^2 = x^2 - 2xy + y^2$$

EXAMPLE 6
Find the following products.

(a) $(x + 7)^2 = x^2 + 2 \cdot x \cdot 7 + 7^2 = x^2 + 14x + 49$

(b) $(y - 8)^2 = y^2 - 2 \cdot y \cdot 8 + 8^2 = y^2 - 16y + 64$

(c) $(2x + 3y)^2 = (2x)^2 + 2 \cdot (2x) \cdot (3y) + (3y)^2 = 4x^2 + 12xy + 9y^2$

(d) $(7a - 5b)^2 = (7a)^2 - 2 \cdot (7a) \cdot (5b) + (5b)^2 = 49a^2 - 70ab + 25b^2$

As the products in the definition of the square of a binomial show,

$$(x + y)^2 \neq x^2 + y^2.$$

More generally,

$$(x + y)^n \neq x^n + y^n.$$

Self-Check 3

Find the following products.

1. $(4x + 9)(4x - 9)$ 2. $(5p + 6q)(5p - 6q)$

3. $(3x + y)^2$ 4. $(7a - 2b)^2$

Self-Check Answers

1.1 $-36x^5y^9$

1.2 $12x^5 - 27x^3$

1.3 $8x^3 - 6x^2 - 27x$

1.4 $6a^2 - ab - b^2$

2.1 $6x^2 + 13x + 5$

2.2 $54y^2 + 33y - 7$

2.3 $18x^2 - 39xy - 7y^2$

2.4 $21a^2 - 43ab + 20b^2$

3.1 $16x^2 - 81$

3.2 $25p^2 - 36q^2$

3.3 $9x^2 + 6xy + y^2$

3.4 $49a^2 - 28ab + 4b^2$

3.3 EXERCISES

Match each product on the left with the correct polynomial on the right in exercises 1-4.

1. $(5x - 2)(8x - 7)$

A. $40x^2 + 51x + 14$

2. $(5x - 2)(8x + 7)$

B. $40x^2 - 51x + 14$

3. $(5x + 2)(8x - 7)$

C. $40x^2 - 19x - 14$

4. $(5x + 2)(8x + 7)$

D. $40x^2 + 19x - 14$

Find each product in exercises 5-34.

5. $(-3x^6)(5x^4)$

6. $(9y^{10})(10y^{12})$

7. $(4x^8y)(-5x^3y^2)$

8. $(-8m^2n^4)(10mn^6)$

9. $-2(4x^3 + 5x^2)$

10. $-7(6t^8 - 2t^4)$

11. $(3x - 2)(2x - 9)$

12. $(5x + 6)(7x - 2)$

13. $(5t - s)(3t - 11s)$

14. $(8r - 5u)(3r + 7u)$

15. $2x + 1$
$\underline{3x - 4}$

16. $5m - 3$
$\underline{2m - 7}$

17. $(3x - 8)(3x + 8)$

18. $(7x + 9y)(7x - 9y)$

19. $m(m + 9)(m - 4)$

20. $p(p - 7)(p - 10)$

21. $3z(2z + 3)(3z - 5)$

22. $3y(4y - 9)(3y - 1)$

23. $x^3(2x + 5)(2x - 5)$

24. $y^2(3y - 7)(3y + 7)$

25. $(2p + 5)(p^2 - p + 5)$

26. $(3z - 8)(z^3 + 2z - 6)$

27. $3x^2 - 2x + 5$
$\underline{4x - 7}$

28. $-y^2 - 4y + 7$
$\underline{3y + 2}$

29. $4x^3 - 3x^2 + x + 2$
$\underline{5x + 1}$

30. $5z^4 + 2z^3 + 3z - 7$
$\underline{2z - 8}$

31. $-x^2 - 5x - 2$
$\underline{2x^3 - 5x}$

32. $5k^2 + 6k + 4$
$\underline{- k^2 + 3k}$

33. $5x^2 + 6x + 1$
$\underline{4x^2 - 2x - 1}$

34. $3y^2 - 3y + 7$
$\underline{-y^2 + 6y - 4}$

35. What type of polynomials can be multiplied by the FOIL method? Describe the method in your own words.

36. Make a list of special product. Give examples with solutions, and explain in your own words how to find these special products.

Find each product in exercises 37-48.

37. $(5p + 2)(5p - 2)$ **38.** $(5x + 6)(5x - 6)$ **39.** $(8x - 1)(8x + 1)$

40. $(3y + 7)(3y - 7)$ **41.** $(7x + 2y)(7x - 2y)$ **42.** $(5r + 2s)(5r - 2s)$

43. $\left(3x - \dfrac{2}{3}\right)\left(3x + \dfrac{2}{3}\right)$ **44.** $\left(2t - \dfrac{5}{4}\right)\left(2t + \dfrac{5}{4}\right)$ **45.** $\left(3m + n^2\right)\left(3m - n^2\right)$

46. $\left(2j^3 + 6k\right)\left(2j^3 - 6k\right)$ **47.** $\left(7y^5 - 3\right)\left(7y^5 + 3\right)$ **48.** $\left(2x^3 - 1\right)\left(2x^3 + 1\right)$

Find each square in exercises 49-56.

49. $(y - 9)^2$ **50.** $(a - 6)^2$ **51.** $(3x + 7)^2$ **52.** $(4y + 1)^2$

53. $(3m - 2n)^2$ **54.** $(4r + 9s)^2$ **55.** $\left(a - \dfrac{2}{3}b\right)^2$ **56.** $\left(x + \dfrac{1}{2}y\right)^2$

57. How do the expressions $(x + y)^2$ and $x^2 + y^2$ differ?

58. Find the product $51 \cdot 49$ using the special product $(x + y)(x - y) = x^2 - y^2$.

Find each product in exercises 59-64.

59. $\left[(2x + 1) + 3y\right]^2$ **60.** $\left[(5m - 2) + n\right]^2$

61. $\left[(4a + b) - 5\right]^2$ **62.** $\left[(4h + k) - 9\right]^2$

63. $\left[(3a + b) - 2\right]\left[(3a + b) + 2\right]$ **64.** $\left[(4x + y) + 5\right]\left[(4x + y) - 5\right]$

Find each product in exercises 65-76.

65. $(2a + b)\left(4a^2 - 3ab + 5b^2\right)$ **66.** $(5x + 2y)\left(3x^2 + 3xy - 4y^2\right)$

67. $(5x - z)\left(x^3 - 5x^2z + 3xz^2 + z^3\right)$ **68.** $(2x + 3y)\left(x^3 - 2x^2y - 8xy^2 + 10y^3\right)$

69. $\left(m^2 - 3mn + n^2\right)\left(m^2 + 3mn - n^2\right)$ **70.** $(4 + x + y)(-4 + x - y)$

71. $ab(a + b)(a - b)(a + 2b)$ **72.** $xy(x - 2y)(x + 2y)(x - 3y)$

73. $(y + 1)^3$ **74.** $(x - 2)^3$ **75.** $(z - 2)^4$ **76.** $(p + 3)^4$

3.4 Introduction to Factoring; Special Factoring

Writing a polynomial as the product of two or more simpler polynomials is called **factoring** the polynomial. For example, the product of $2x$ and $7x - 3$ is $14x^2 - 6x$, and $14x^2 - 6x$ can be factored as the product $2x(7x - 3)$.

$$2x(7x - 3) = 14x^2 - 6x \qquad \text{Multiplication}$$

$$14x^2 - 6x = 2x(7x - 3) \qquad \text{Factoring}$$

Notice that both multiplication and factoring are examples of the distributive property, used in opposite directions. Factoring is the reverse of multiplying.

The Greatest Common Factor: The first step in factoring a polynomial is to find the *greatest common factor* for the terms of the polynomial. The **greatest common factor (GCF)** is the largest term that is a factor of all terms in the polynomial. For example, the greatest common factor for $6x + 8$ is 2, since 2 is the largest number that is a factor of both $6x$ and 8. Using the distributive property,

$$6x + 8 = 2(3x) + 2(4) = 2(3x + 4)$$

As a check, multiply 2 and $3x + 4$. The result should be $6x + 8$. This process is called **factoring out the greatest common factor**.

EXAMPLE 1
Factor out the greatest common factor.

(a) $36x - 24$
Since 12 is the greatest common factor, factor 12 from each term.
$$36x - 24 = 12(3x) - 12 \cdot 2 = 12(3x - 2)$$

(b) $12x + 16y = 4(3x + 4y)$

(c) $5x + 6$ There is no common factor other than 1.

(d) $14 + 28z = 14 \cdot 1 + 14(2z) = 14(1 + 2z)$ \qquad 14 is the GCF.

(e) $10x^2 + 15x^3 = 5x^2(2) + 5x^2(3x) = 5x^2(2 + 3x)$ \qquad $5x^2$ is the GCF.

When the coefficient of the term of greatest degree is negative, it is sometimes preferable to factor out the -1 that is understood along with the greatest common factor.

EXAMPLE 2
Factor $-x^4 + 5x^2 - 2x$ in two ways.
First, x could be used as a common factor, giving
$$-x^4 + 5x^2 - 2x = x\left(-x^3\right) + x(5x) + x(-2)$$
$$= x\left(-x^3 + 5x - 2\right).$$

Alternatively, because of the leading negative sign, $-x$ could be used as the common factor.

$$-x^4 + 5x^2 - 2x = -x\left(x^3\right) + (-x)(-5x) + (-x)(2)$$
$$= -x\left(x^3 - 5x + 2\right).$$

Either answer is correct.

Self-Check 1

Factor.

1. $25x^2 - 15x$ **2.** $12mn + 36m$

Factor in two ways.

3. $-x^5 + 3x^4 - 5x^2$ **4.** $a^3 - 3a^2 + a$

Factor by Grouping: Sometimes a polynomial has a greatest common factor of 1, but it still may be possible to factor the polynomial by using a process called **factoring by grouping**. We usually factor by grouping when a polynomial has more than three terms.

For example, to factor the polynomial

$$ax - bx + ay - by,$$

group the terms as follows.

Terms with common factors

$$\downarrow \qquad\qquad \downarrow$$

$$(ax - bx) + (ay - by)$$

Then factor $ax - bx$ as $x(a - b)$ and factor $ay - by$ as $y(a - b)$ to get

$$ax - bx + ay - by = x(a - b) + y(a - b).$$

On the right, the common factor is $a - b$. The final factored form is

$$ax - bx + ay - by = (a - b)(x + y).$$

The steps used in factoring by grouping are listed below.

Rule: Factoring by Grouping

Step 1   **Group terms.** Collect the terms into groups so that each group has a common factor.

Step 2   **Factor within the groups.** Factor out the common factor in each group.

Step 3   **Factor the entire polynomial.** If each group now has a common factor, factor it out. If not, try a different grouping.

EXAMPLE 3

Factor $5a - 5b - ay + by$.

Grouping terms as above gives

$$(5a - 5b) + (-ay + by),$$

or

$$5(a - b) + y(-a + b).$$

There is no simple common factor here. However, if we factor out $-y$ instead of y in the second group of terms, we get

$$5(a - b) - y(a - b) = (a - b)(5 - y).$$

Check by multiplying.

EXAMPLE 4

Factor $8x^2 - 6x - 20x + 15$ by grouping.

Note that we must factor -5 rather than 5 from the second group in order to get a common factor of $4x - 3$.

$$\left(8x^2 - 6x\right) + \left(-20x + 15\right) = 2x\left(4x - 3\right) - 5\left(4x - 3\right) \quad \text{Factor out } 2x \text{ and } -5.$$
$$= \left(4x - 3\right)\left(2x - 5\right) \qquad \text{Factor out } \left(4x - 3\right).$$

Self-Check 2

Factor.

1. $3x - 3y + 7x^2 - 7xy$

2. $20x^2 - 4x + 15x - 3$

3. $18y^2 + 27y - 8y - 12$

4. $3ax + 6bx - 5ay - 10by$

The Difference of Two Squares: The product of the sum and difference of two terms leads to a **difference of two squares**, a pattern that occurs often when factoring.

Rule: Difference of Two Squares

$$x^2 - y^2 = \left(x - y\right)\left(x + y\right)$$

EXAMPLE 5

Factor each difference of two squares.

(a) $36x^2 - 49y^2 = \left(6x\right)^2 - \left(7y\right)^2 = \left(6x - 7y\right)\left(6x + 7y\right)$

(b) $64m^2 - 81n^2 = \left(8m\right)^2 - \left(9n\right)^2 = \left(8m - 9n\right)\left(8m + 9n\right)$

(c) $\left(x - 5y\right)^2 - 1 = \left(x - 5y\right)^2 - 1^2$
$$= \left[\left(x - 5y\right) - 1\right]\left[\left(x - 5y\right) + 1\right]$$
$$= \left(x - 5y - 1\right)\left(x - 5y + 1\right)$$

Assuming no greatest common factor (except 1), it is *not* possible to factor (with real numbers) a *sum* of two squares such as $x^2 + 36$. In particular, $x^2 + y^2 \neq \left(x + y\right)^2$.

A Perfect Square Trinomial: When a binomial is squared,

$$\left(x + y\right)^2 = \left(x + y\right)\left(x + y\right)$$

by the distribution property, the result is

$$x^2 + 2xy + y^2.$$

Similarly, $\left(x - y\right)^2 = x^2 - 2xy + y^2$. Each of these trinomials is called a **perfect square trinomial**.

Rule: Perfect Square Trinomials

$$x^2 + 2xy + y^2 = \left(x + y\right)^2 \quad \text{and} \quad x^2 - 2xy + y^2 = \left(x - y\right)^2$$

In this pattern, both the first and the last terms of the trinomial must be perfect squares. In the factored form, twice the product of the first and the last terms must give the middle term of the trinomial. This pattern occurs with many difference symbols (other than x and y).

$$16m^2 - 40m + 25 \qquad\qquad\qquad p^2 - 5p + 16$$

Square trinomial $\qquad\qquad$ Not a squared trinomial; middle term should be $-8p$.

EXAMPLE 6
Factor each of the following.

(a) $16m^2 - 40m + 25$

Here $16m^2 = (4m)^2$ and $25 = 5^2$. The sign on the middle term is $-$, so if $16m^2 - 40m + 25$ is a perfect square trinomial, it will have to be $(4m - 5)^2$.

Take twice the product of the two terms to see if this is correct. We have

$$2(4m)(-5) = -40m,$$

which is the middle term of the given trinomial. Thus,

$$16m^2 - 40m + 25 = (4m - 5)^2.$$

(b) $100x^2 + 60x + 9 = (10x)^2 + 2(10x) \cdot 3 + 3^2 = (10x + 3)^2$

Self-Check 3

Factor.

1. $49a^2 - 16b^2$ $\qquad\qquad$ **2.** $(5m - 1)^2 - 4$

3. $49x^2 - 84x + 36$ $\qquad\qquad$ **4.** $121a^2 + 264ab + 144b^2$

The Difference of Cubes: A difference of two cubes, $x^3 - y^3$, can be factored as follows.

Rule: Difference of Two Cubes

$$x^3 - y^3 = (x - y)(x^2 + xy + y^2)$$

We could check this pattern by finding the product of $x - y$ and $x^2 + xy + y^2$.

EXAMPLE 7
Factor each difference of cubes.

(a) $m^3 - 27 = m^3 - 3^3 = (m - 3)(m^2 + 3m + 9)$

Check: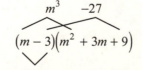

$-3m \longleftarrow$ Opposite of the product of the cube roots is the middle term.

(b) $64x^3 - 27y^3 = (4x)^3 - (3y)^3 = (4x - 3y)(16x^2 + 12xy + 9y^2)$

(c) $1000s^3 - 343t^3 = (10s)^3 - (7t)^3 = (10s - 7t)\left[(10s)^2 + (10s)(7t) + (7t)^2\right]$

$$= (10s - 7t)\left(100s^2 + 70st + 49t^2\right)$$

The Sum of Two Cubes: An expression of the form $x^2 + y^2$ (a sum of two squares) cannot be factored with real numbers. However, a **sum of two cubes** is factored as follows.

Rule: Sum of Two Cubes

$$x^3 + y^3 = (x + y)\left(x^2 - xy + y^2\right)$$

To verify this result, find the product of $x + y$ and $x^2 - xy + y^2$.

Notice that the sign of the second term in the binomial factor of a sum or difference of cubes is *always the same* as the sign in the original polynomial. In the trinomial factor, the first and last terms are *always positive*; the sign of the middle term is *the opposite of* the sign of the second term in the binomial factor.

EXAMPLE 8
Factor each sum of cubes.

(a) $w^3 + 8 = w^3 + 2^3 = (w + 2)(w^2 - 2w + 4)$

(b) $64z^3 + 125 = (4z)^3 + 5^3 = (4z + 5)(16z^2 - 20z + 25)$

(c) $343s^9 + 125t^6 = (7s^3)^3 + (5t^2)^3 = (7s^3 + 5t^2)\left[(7s^3)^2 - (7s^3)(5t^2) + (5t^2)^2\right]$

$$= (7s^3 + 5t^2)(49s^6 - 35s^3t^2 + 25t^4)$$

Self-Check 4

Factor.

1. $x^3 - 64$ 2. $216x^3 - y^6$ 3. $8a^6 + 27$ 4. $y^9 + 1000z^6$

The factoring techniques covered in this section are summarized here.

Factoring out the greatest common factor	$ax^2 - 2a = a(x^2 - 2)$
Factoring by grouping	$ax - ay + bx - by = a(x - y) + b(x - y)$ $= (x - y)(a + b)$
Difference of two square	$x^2 - y^2 = (x - y)(x + y)$
Perfect square trinomial	$x^2 + 2xy + y^2 = (x + y)^2$ $x^2 - 2xy + y^2 = (x - y)^2$
Difference of two cubes	$x^3 - y^3 = (x - y)(x^2 + xy + y^2)$
Sum of two cubes	$x^3 + y^3 = (x + y)(x^2 - xy + y^2)$

Self-Check Answers

1.1 $5x(5x - 3)$ **1.2** $12m(n + 3)$

1.3 $-x^2(x^3 + 3x^2 - 5)$ and $x^2(-x^3 - 3x^2 + 5)$

1.4 $a(a^2 - 3a + 1)$ and $-a(-a^2 + 3a - 1)$

2.1 $(x - y)(3 + 7x)$ **2.2** $(5x - 1)(4x + 3)$

2.3 $(2y + 3)(9y - 4)$ **2.4** $(a + 2b)(3x - 5y)$

3.1 $(7a - 4b)(7a + 4b)$ **3.2** $(5m - 3)(5m + 1)$

3.3 $(7x - 6)^2$ **3.4** $(11a + 12b)^2$

4.1 $(x - 4)(x^2 + 4x + 16)$ **4.2** $(6x - y^2)(36x^2 + 6xy^2 + y^4)$

4.3 $(2a^2 + 3)(4a^4 - 6a^2 + 9)$ **4.4** $(y^3 + 10z^2)(y^6 - 10y^3z^2 + 100z^4)$

3.4 EXERCISES

1. Explain in your own words what it means to factor a polynomial. Discuss the methods of factoring presented in this section, and include an example of each.

2. Write a paragraph explaining how multiplication of polynomials and factoring are related. Give an example.

3. What is the greatest common factor of the terms $3x^6(x + 1)^7$ and $12x^3(x + 1)^{10}$?

4. Why is $(3x + 4)(5x - 7) + (3x + 4)(6x - 1)$ not in factored form?

Factor out the greatest common factor in exercises 5-29. Simplify the factors, if possible.

5. $5c + 25$ 6. $8p - 26$ 7. $7j^3 + 21j$

8. $6z^4 + 60z$ 9. $st - 9st^2$ 10. $8g^2h + gh$

11. $-6p^2q^5 - 3p^3q^2$ 12. $-8a^7b^3 - 24a^6b^4$

13. $15x^8 - 20x^5 + 25x^2$ 14. $8z^5 - 24z^7 - 48z^3$

15. $6x^2y^3 - 36x^8y^4 + 42x^3y^3$ 16. $14c^3d^9 - 7c^2d^8 + 21cd^5$

17. $-14x^2y + 7x^3y^7 - 21x^3y^2$ 18. $-40u^4v^7 + 60u^3v^5 + 80u^2v^4$

19. $(b - 2)(b + 9) + (b - 2)(2b + 3)$ 20. $(z - 5)(z + 3) + (z - 5)(3z + 1)$

21. $(3x - 2)(x + 5) - (3x - 2)(8x - 7)$ 22. $(4y - 7)(y + 6) - (4y - 7)(2y - 1)$

23. $-y^7(a + b) - y^8(c + d)$ 24. $-z^3(r + s) - z^4(t + u)$

25. $3(3 - x)^2 - 4(3 - x)^3 + 5(3 - x)$ 26. $2(6 - x)^4 - 3(6 - x)^3 + 4(6 - x)^2$

27. $5(2 - x)^2 - (2 - x)^3 - 2(2 - x)$ 28. $3(a - b) + 2(a - b)^3 - (a - b)^2$

29. $4(2z + 1)^2 + 2(2z + 1)^4 - (2z + 1)^3$

30. Which of the following factored forms of

$$8x^3y^5 - 16x^6y^2 + 32x^5y^8$$

has the greatest common factor as one of its factors?

(a) $8xy\left(x^2y^4 - 2x^5y + 4x^4y^7\right)$ (b) $8x^3y^2\left(y^3 - 2x^3 + 4x^2y^6\right)$

(c) $2x^3y^2\left(4y^3 - 8x^3 + 16x^2y^6\right)$ (d) $8x^2y^2\left(xy^3 - 2x^4 + 4x^3y^6\right)$

In exercises 31-34, factor each polynomial twice. First, use a common factor with a positive coefficient, and then use a common factor with a negative coefficient.

31. $-3x^5 + 6x^4 + 9x^3$ **32.** $-7y^3 + 21y^4 - 14y^5$

33. $-36y^4z^5 - 12y^2z^4 + 24y^5z^3$ **34.** $-72a^2b^3 - 24a^3b^2 - 36a^3b^5$

Factor by grouping in exercises 35-52.

35. $ab + 4b^2 + ad + 4bd$ **36.** $3x + 3y + xh + yh$

37. $3k + 6km + 7p + 14mp$ **38.** $4xy + 4xz + 5yz + 5z^2$

39. $x^2 - 3x - 12 + 4x$ **40.** $z^2 - 5z - 35 + 7z$

41. $a^2 - 3bc + ac - 3ab$ **42.** $x^2 - 9yz + 3xz - 3xy$

43. $3x^2 + 12x - 20 - 5x$ **44.** $6y + 5y^2 - 10y - 12$

45. $-20x^2 + 4xy - 25xy + 5y^2$ **46.** $-7a^2 + 2ab - 14ab + 4b^2$

47. $6 + ab - 3b - 2a$ **48.** $2xy^2 - 5 - 10y^2 + x$

49. $7x^4 + 2x^3 - 14x - 4$ **50.** $a^3b - 5 - 5b + a^3$

51. $1 - x + 4xy - 4y$ **52.** $3s^2t + 6s^2 + t + 2$

53. Match each binomial with its correct form.

(a) $x^2 - y^2$ A. $(x - y)(x^2 + xy + y^2)$
(b) $y^2 - x^2$ B. $(x - y)(x + y)$
(c) $x^3 - y^3$ C. $(x + y)(x^2 - xy + y^2)$
(d) $x^3 + y^3$ D. $(y - x)(y + x)$

54. Match each trinomial with its correct form.

(a) $x^2 + 2xy + y^2$ A. $(x^2 + y^2)^2$
(b) $x^2 - 2xy + y^2$ B. $(x - y)^2$
(c) $x^4 + 2x^2y^2 + y^4$ C. $(x + y)^2$
(d) $x^4 - 2x^2y^2 + y^4$ D. $(x - y)^2(x + y)^2$

55. Which of the following may be considered a sum or difference of cubes?

(a) $27 - y^3$ (b) $125 - z^9$ (c) $x^3 + y^4$ (d) $(x + y)^3 + 1$

56. Which of the following may be considered a perfect square trinomial?

(a) $a^2 - 10a + 25$ (b) $4s^2 - 12s + 9$
(c) $9z^4 + 42z^2 + 49$ (d) $100x^2 - 53x + 81$

Factor each polynomial in exercises 57-92.

57. $x^2 - 16$ **58.** $y^2 - 25$ **59.** $36x^2 - 49$

60. $81y^2 - 4$ **61.** $25x^2 - 36y^2$ **62.** $81a^2 - 49b^2$

63. $3x^4 - 243y^4$ **64.** $64a^4 - 4b^4$ **65.** $(x + y)^2 - 16$

66. $(a + b)^2 - 64$ **67.** $49 - (x + 2y)^2$ **68.** $81 - (a + 3b)^2$

69. $(a + b)^2 - (a - b)^2$ **70.** $(s + t)^2 - (s - t)^2$ **71.** $k^2 - 8k + 16$

72. $x^2 + 2x + 1$ **73.** $9a^2 - 6ab + b^2$ **74.** $25s^2 + 10st + t^2$

75. $49x^2 - 14x + 1 - y^2$ **76.** $25m^2 + 20m + 4 - n^2$ **77.** $16x^2 - 24x + 9 - y^2$

78. $16x^2 + 40x + 25 - y^2$ **79.** $6x^2 + 36x + 54$ **80.** $5y^2 - 90y + 405$

81. $(a + b)^2 - 12(a + b) + 36$ **82.** $(x - y)^2 + 16(x - y) + 64$

83. $64a^3 + b^3$ **84.** $8x^3 - 27y^3$

85. $1000x^3 - 27y^3$ **86.** $343s^3 + 8t^3$

87. $81a^6 - 1$ **88.** $64a^9 + 27$

89. $(x + y)^3 - 1$ **90.** $(a + b)^3 + 8$

91. $729a^6 - 512b^{15}$ **92.** $125x^9 + y^{12}$

3.5 | Factoring Trinomials

When the Coefficient of the Squared Term is 1: Let's begin by comparing multiplication and factoring.

Multiplication
$$\text{Factored Form} \rightarrow \quad (x + 2)(x - 6) = x^2 - 4x - 12 \quad \leftarrow \text{Product}$$
Factoring

Since multiplying and factoring are operations that "undo" each other, factoring trinomials involves using FOIL backwards. As shown next, the x^2-term came from multiplying x and x, and -12 came from multiplying 2 and -6.

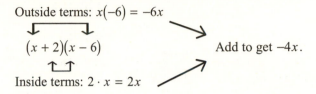

$$\text{Product of } x \text{ and } x \quad \text{is } x^2$$
$$\downarrow \qquad \downarrow \qquad \downarrow$$
$$(x + 2)(x - 6) = x^2 - 4x - 12$$
$$\uparrow \qquad \uparrow \qquad\qquad \uparrow$$
$$\text{Product of 2 and } -6 \qquad \text{is } -12.$$

We found the $-4x$ in $x^2 - 4x - 12$ by multiplying the outside terms, and then the inside terms, and adding.

Outside terms: $x(-6) = -6x$

$(x + 2)(x - 6)$ Add to get $-4x$.

Inside terms: $2 \cdot x = 2x$

Based on this example, follow these steps to factor a trinomial $x^2 + bx + c$, with 1 as the coefficient of the squared term.

Procedure: Factoring $x^2 + bx + c$

Step 1 **Find pairs whose product is c.** Find all pairs of integers whose product is the third term of the trinomial (c).

Step 2 **Find pairs whose sum is b.** Choose the pair whose sum is the coefficient of the middle term (b).

If there are no such integers, the polynomial cannot be factored.

A polynomial that cannot be factored with integer coefficients is **prime**.

EXAMPLE 1
Factor each polynomial.

(a) $x^2 + 3x - 18$

Step 1 Find pairs of numbers whose product is -18.

Step 2 Write sums of those numbers.

$1(-18)$	$1 + (-18) = -17$
$-1(18)$	$-1 + 18 = 17$
$2(-9)$	$2 + (-9) = -7$
$-2(9)$	$-2 + 9 = 7$
$3(-6)$	$3 + (-6) = -3$
$-3(6)$	$-3 + 6 = 3$ ← Coefficient of middle term

The required numbers are -3 and 6.

$$x^2 + 3x - 18 = (x - 3)(x + 6)$$

Check by finding the product of $x - 3$ and $x + 6$.

(b) $z^2 + 10z + 16$

Look for two numbers with a product of 16 and a sum of 10. Of all pairs of numbers having a product of 16, only the pair 8 and 2 has a sum of 10. Therefore,

$$z^2 + 10z + 16 = (z + 8)(z + 2).$$

Because of the commutative property, it would be equally correct to write $(z + 2)(z + 8)$.
Check by multiplying.

EXAMPLE 2

Factor $m^2 + 4m + 5$.

Look for two numbers whose product is 5 and whose sum is 4. Only two pairs of integers, 5 and 1 and −5 and −1, give a product of 5. Neither of these pairs has a sum of 4, so $m^2 + 4m + 5$ cannot be factored with integer coefficients, and is prime.

Factoring a trinomial that has more than one variable uses a similar process, as shown in the next example.

EXAMPLE 3

Factor $a^2 + 2ab - 8b^2$.

Look for two expressions whose product is $-8b^2$ and whose sum is $2b$. The quantities $4b$ and $-2b$ have the necessary product and sum so

$$a^2 + 2ab - 8b^2 = (a + 4b)(a - 2b).$$

Sometimes a trinomial has a common factor that should be factored out first.

EXAMPLE 4

Factor $2z^3 + 4z^2 - 96z$.

Start by factoring out the greatest common factor, $2z$.

$$2z^3 + 4z^2 - 96z = 2z(z^2 + 2z - 48)$$

Now factor $z^2 + 2z - 48$, look for two integers whose product is −48 and whose sum is 2. The necessary integers are −6 and 8, with

$$2z^3 + 4z^2 - 96z = 2z(z - 6)(z + 8).$$

Self-Check 1

Factor.

1. $p^2 - 8p + 15$ 2. $x^2 - 3xy - 28y^2$

3. $r^2 - 2r + 8$ 4. $3x^3 - 21x^2 - 54x$

When the Coefficient of the Squared Term is Not 1: Suppose we want to factor a trinomial of the form $ax^2 + bx + c$, where $a \neq 1$. To factor $2x^2 + 7x + 3$, for example, first identify the values of a, b, and c.

$$ax^2 + bx + c$$
$$\downarrow \quad\quad \downarrow \quad\quad \downarrow$$
$$2x^2 + 7x + 3$$

The product ac is $2 \cdot 3 = 6$, so we must find integers having a product of 6 and a sum of 7 (since the middle term has coefficient 7). The necessary integers are 1 and 6, so we write $7x$ as $1x + 6x$, or $x + 6x$, giving

$$2x^2 + 7x + 3 = 2x^2 + x + 6x + 3.$$

$$x + 6x = 7x$$

Now we factor by grouping.

$$2x^2 + x + 6x + 3 = x(2x + 1) + 3(2x + 1)$$
$$2x^2 + 7x + 3 = (2x + 1)(x + 3)$$

EXAMPLE 5

Factor $15d^2 + 4d - 3$.

Since $a = 15$, $b = 4$, and $c = -3$, the product ac is -45. The two integers whose product is -45 and whose sum is 4 are 9 and -5.

$$15d^2 + 4d - 3 = 15d^2 + 9d - 5d - 3 \qquad \text{Write } 4d \text{ as } 9d - 5d.$$
$$= 3d(5d + 3) - 1(5d + 3) \qquad \text{Factor by grouping.}$$
$$15d^2 + 4d - 3 = (5d + 3)(3d - 1) \qquad \text{Factor out the common factor.}$$

There is another approach to factoring trinomials in $ax^2 + bx + c$ form. Let's factor $7x^2 + 10x + 3$ as an example.

To factor this polynomial, find the correct numbers to put in the blanks.

$$7x^2 + 10x + 3 = (\underline{}x + \underline{})(\underline{}x + \underline{})$$

Addition signs are used since all the signs in the polynomial indicate addition. The first two expressions have a product of $7x^2$, so they must be $7x$ and x.

$$7x^2 + 10x + 3 = (7x + \underline{})(x + \underline{})$$

The product of the two last terms must be 3, so the numbers must be 3 and 1. There is a choice. The 3 could be used with the $7x$ or with the x. Only one of these choices can give the correct middle term, $10x$. Use FOIL to try each one.

$21x$	$7x$
$(7x + 1)(x + 3)$	$(7x + 3)(x + 1)$
x	$3x$
$21x + x = 22x$	$7x + 3x = 10x$
Wrong middle term	Correct middle term

Therefore, $7x^2 + 10x + 3 = (7x + 3)(x + 1)$.

This alternative method of factoring a trinomial $ax^2 + bx + c$, $a \neq 1$, is summarized below.

Procedure: Factoring $ax^2 + bx + c$

Step 1   **Find pairs whose product is *a*.** Write all pairs of integer factors of the coefficient of the squared term (a).

Step 2   **Find pairs whose product is *c*.** Write all pairs of integer factors of the last term (c).

Step 3   **Choose inner and outer terms.** Use FOIL and various combinations of the factors from Steps 1 and 2 until the necessary middle term is found.

If no such combinations exist, the polynomial is prime.

It takes a great deal of practice using different methods to become proficient at factoring.

EXAMPLE 6

Factor $24a^2 + 7ab - 6b^2$.

There is no common factor (except 1). Go through the steps to factor the trinomial. There are many possible factors of both 24 and −6. Let's try 6 and 4 for 24 and −2 and 3 for −6.

$$(6a - 2b)(4a + 3b) \qquad\qquad (6a + 3b)(4a - 2b)$$

Wrong: common factor   Wrong: common factors

Since 6 and 4 do not work as factors of 24, try 8 and 3 instead, with −2 and 3 as factors of −6.

$$\underset{\text{Wrong: common factors}}{(8a - 2b)(3a + 3b)} \qquad\qquad \overset{-16ab}{\underset{9ab}{(8a + 3b)(3a - 2b)}}$$

$$-16ab + 9ab = -7ab$$

The result on the right differs from the correct middle term only in sign, so exchange the signs in the factors. Check by multiplying.

$$24a^2 + 7ab - 6b^2 = (8a - 3b)(3a + 2b)$$

EXAMPLE 7

Factor $-20k^2 + 7k + 3$.

While it is possible to factor this polynomial directly, it is helpful to first factor out −1. Then proceed as in the earlier examples.

$$-20k^2 + 7k + 3 = -1(20k^2 - 7k - 3) = -1(4k + 1)(5k - 3)$$

Self-Check 2

Factor.

1. $10x^2 - x - 2$ 2. $35p^2 - 11p - 6$

3. $6x^2 + 37xy + 45y^2$ 4. $-40x^2 + 27x + 4$

Self-Check Answers

1.1 $(p - 5)(p - 3)$ 1.2 $(x - 7y)(x + 4y)$ 1.3 prime   1.4 $3x(x - 9)(x + 2)$

2.1 $(5x + 2)(2x - 1)$ 2.2 $(7p + 2)(5p - 3)$

2.3 $(3x + 5y)(2x + 9y)$ 2.4 $-1(8x + 1)(5x - 4)$

3.5 EXERCISES

1.   Match each polynomial in Column I with the correct form from Column II.

	I	II
(a)	$x^2 - 5x - 6$	A. $(x + 3)(x + 2)$
(b)	$x^2 + 5x - 6$	B. $(x - 6)(x + 1)$
(c)	$x^2 + 5x + 6$	C. $(3x - 4)(5x + 9)$
(d)	$x^2 - 5x + 6$	D. $(x - 3)(x - 2)$
(e)	$15x^2 + 7x - 36$	E. $(3x + 4)(5x - 9)$
(f)	$15x^2 - 7x - 36$	F. $(x + 6)(x - 1)$

2. Explain in your own words how when working Exercise 1, you can immediately narrow down your choices to two factored forms in the column on the right.

Factor each trinomial in exercises 3-52.

3. $y^2 + 14y + 24$

4. $m^2 + 10m + 21$

5. $r^2 - 11r + 30$

6. $x^2 - 13x + 22$

7. $y^2 + 2y - 24$

8. $y^2 - 5y - 14$

9. $a^2b^2 + 6ab + 5$

10. $x^2y^2 + 8xy + 7$

11. $r^2 - 10rs + 16s^2$

12. $u^2 - 11uv + 18v^2$

13. $s^2 + 7st - 18t^2$

14. $m^2 - 3mn - 18n^2$

15. $15y^2 + 47y + 36$

16. $6z^2 + 31z + 35$

17. $12x^2 - 44x - 45$

18. $63p^2 + 11p - 2$

19. $15j^2 + 22j - 48$

20. $15y^2 + 37y + 20$

21. $15a^2 - 26ab + 8b^2$

22. $6x^2 - 19xy + 8y^2$

23. $49r^2 + 42rs + 9s^2$

24. $36x^2 - 12xy + y^2$

25. $49c^2 - 14cd + d^2$

26. $100u^2 + 60uv + 9v^2$

27. $6g^2 + 7gh - 5h^2$

28. $6m^2 - 7mn + 2n^2$

29. $2p^2 - 26p + 44$

30. $10d^2 + 140d + 240$

31. $6a^2 - 78a + 132$

32. $12j^2 + 120j + 252$

33. $x^3 - x^2 - 12x$

34. $y^3 - 3y^2 - 18y$

35. $3x^4 + 18x^3 + 15x^2$

36. $5a^2b + 40ab + 35b$

37. $12x^3y + 38x^2y + 16xy$

38. $45p^2z^3 + 141pz^3 + 108z^3$

39. $200x^2y^2 + 120xy^2 + 18y^2$

40. $72au^2 + 24auv + 2av^2$

41. $48x^2y - 176xy^2 - 180y^3$

42. $315p^4q - 55p^3q - 10p^2q$

43. $30a^2z - 52acz + 16c^2z$

44. $4x^3 - 24x^2 - 64x$

45. $90p^3 + 120p^2 + 40p$

46. $13z^5 - 39z^4 - 52z^3$

47. $-x^2 - 3x + 54$

48. $-y^2 + 8y + 33$

49. $-10k^2 - 3k + 18$ **50.** $-15v^2 + 43v - 8$

51. $-70x^3 - 32x^2 + 6x$ **52.** $-54xy^3 - 27xy^2 + 6xy$

53. When a student was given the polynomial $27x^2 - 9x - 6$ to factor completely on a test, the student lost some credit when her answer was $(9x + 3)(3x - 2)$. She complained to her teacher that when we multiply $(9x + 3)(3x - 2)$ we get the original polynomial. Write as short explanation of why she lost some credit for her answer, even though the product is indeed $27x^2 - 9x - 6$.

54. Write an explanation as to why most people would find it more difficult to factor $42x^2 + 34x - 16$ than $43x^2 + 342x - 16$.

3.6 | Division of Polynomials

Dividing a Polynomial by a Monomial: Earlier we added, subtracted, and multiplied polynomials. Now we discuss polynomial division, beginning with division by a monomial. To divide a polynomial by a monomial, divide each term in the polynomial by the monomial, and then write each quotient in lowest terms.

EXAMPLE 1

Divide $20x^2 - 6x - 4$ by 2.

Divide each term of the polynomial by 2. Then write the result in lowest terms.

$$\frac{20x^2 - 6x - 4}{2} = \frac{20x^2}{2} - \frac{6x}{2} - \frac{24}{2}$$
$$= 10x^2 - 3x - 12$$

Check this answer by multiplying it by the divisor, 2. If you are correct, you should get $20x^2 - 6x - 4$ as the result.

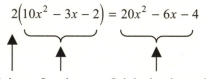

$$2(10x^2 - 3x - 2) = 20x^2 - 6x - 4$$

Divisor   Quotient   Original polynomial

EXAMPLE 2

Find each quotient.

(a) $\dfrac{26m^5 - 10m^3 - 8m}{4m^3} = \dfrac{26m^5}{4m^3} - \dfrac{10m^3}{4m^3} - \dfrac{8m}{4m^3}$ Divide each term by $4m^3$.

        $= \dfrac{13m^2}{2} - \dfrac{5}{2} - \dfrac{2}{m^2}$ Write in lowest terms.

This result is not a polynomial; the quotient of two polynomials need not be a polynomial.

(b) $\dfrac{9x^2y^2 + 5xy^2 - 4xy}{x^2y^2} = \dfrac{9x^2y^2}{x^2y^2} + \dfrac{5xy^2}{x^2y^2} - \dfrac{4xy}{x^2y^2}$

        $= 9 + \dfrac{5}{x} - \dfrac{4}{xy}$

Self-Check 1

Divide.

1. $\dfrac{35x^2 - 10x + 15}{5}$

2. $\dfrac{32d^6 + 16d^5 - 8d}{4d^3}$

3. $\dfrac{18f^4 - 25f^3 - 9f}{6f^2}$

4. $\dfrac{10a^3b^3 + 5a^2b^2 - 7ab}{5a^2b^2}$

Dividing a Polynomial by a Polynomial: The quotient of two polynomials can be found by factoring and then dividing out any common factors. For instance,

$$\frac{6x^2 + 19x - 7}{6x - 2} = \frac{(3x - 1)(2x + 7)}{2(3x - 1)} = \frac{2x + 7}{2}.$$

When the polynomials in a quotient of two polynomials have no common factors or cannot be factored, they can be divided by a process very similar to that for dividing one whole number by another. The following examples show how this is done.

EXAMPLE 3

Divide $2m^2 - m - 6$ by $m - 2$.

Write the problem making sure that both polynomials are written in descending powers of the variables.

$$m - 2\overline{\smash{)}2m^2 - m - 6}$$

Divide the first term of $2m^2 - m - 6$ by the first term of $m - 2$. Since $\dfrac{2m^2}{m} = 2m$, place this result above the division line.

$$\begin{array}{r} 2m \\ m - 2\overline{\smash{)}2m^2 - m - 6} \end{array} \quad \leftarrow \text{Result of } \tfrac{2m^2}{m}$$

Multiply $m - 2$ and $2m$, and write the result below $2m^2 - m - 6$.

$$\begin{array}{r} 2m \\ m - 2\overline{\smash{)}2m^2 - m - 6} \\ \underline{2m^2 - 4m} \end{array} \quad \leftarrow 2m(m - 2) = 2m^2 - 4m$$

Now subtract $2m^2 - 4m$ from $2m^2 - m - 6$. Do this by mentally changing the signs on $2m^2 - 4m$ and *adding*.

$$\begin{array}{r} 2m \\ m - 2\overline{\smash{)}2m^2 - m - 6} \\ \underline{2m^2 - 4m} \\ 3m \leftarrow \text{Subtract.} \end{array}$$

Bring down -6 and continue by dividing $3m$ by m.

$$
\begin{array}{r}
2m + 3 \quad \leftarrow \frac{3m}{m} = 3 \\
m - 2 \overline{\smash{\big)}\, 2m^2 - m - 6} \\
\underline{2m^2 - 4m} \\
3m - 6 \quad \leftarrow \text{Bring down } -6. \\
\underline{3m - 6} \quad \leftarrow 3(m - 2) = 3m - 6 \\
0 \quad \leftarrow \text{Subtract.}
\end{array}
$$

Finally, $\dfrac{2m^2 - m - 6}{m - 2} = 2m + 3$.   Check by multiplying $m - 2$ and $2m + 3$.   The result should be $2m^2 - m - 6$.   Since there is no remainder, this quotient could have been found by factoring and canceling the common factor of $m - 2$.

Self-Check 2

Divide.

1. $\dfrac{3x^2 - 7x - 20}{x - 4}$ 2. $\dfrac{12x^3 + x^2 - 57x + 14}{x - 2}$

Sometimes the polynomial has a "missing" term.   In that case, add a term as a placeholder.

EXAMPLE 4

Divide $4x^3 - 3x + 6$ by $x - 2$.

Make sure that $4x^3 - 3x + 6$ is in descending powers of the variable.   Add a term with 0 coefficient as a placeholder for the missing x^2 term.

$$
\overset{\text{Missing term}}{\downarrow}
$$

$$
x - 2 \overline{\smash{\big)}\, 4x^3 + 0x^2 - 3x + 6}
$$

Start with $\dfrac{4x^3}{x} = 4x^2$.

$$
\begin{array}{r}
4x^2 \quad \leftarrow \frac{4x^3}{x} = 4x^2 \\
x - 2 \overline{\smash{\big)}\, 4x^3 + 0x^2 - 3x + 6} \\
\underline{4x^3 - 8x^2} \quad \leftarrow 4x^2(x - 2)
\end{array}
$$

Subtract by changing the signs on $4x^3 - 8x^2$ and adding.

$$
\begin{array}{r}
4x^2 \\
x - 2 \overline{\smash{\big)}\, 4x^3 + 0x^2 - 3x + 6} \\
\underline{4x^3 - 8x^2} \\
8x^2 \quad \leftarrow \text{Subtract.}
\end{array}
$$

Bring down the next term.

$$
\begin{array}{r}
4x^2 \\
x - 2\overline{\smash{)}4x^3 + 0x^2 - 3x + 6} \\
\underline{4x^3 - 8x^2} \\
8x^2 - 3x
\end{array}
$$

$\leftarrow$ Bring down $-3x$.

In the next step, $\dfrac{8x^2}{x} = 8x$.

$$
\begin{array}{r}
4x^2 + 8x \\
x - 2\overline{\smash{)}4x^3 + 0x^2 - 3x + 6} \\
\underline{4x^3 - 8x^2} \\
8x^2 - 3x \\
\underline{8x^2 - 16x} \\
13x + 6
\end{array}
$$

$\leftarrow \frac{8x^2}{x} = 8x$

$\leftarrow 8x(x - 2)$

$\leftarrow$ Subtract and bring down 6.

Finally, $\dfrac{13x}{x} = 13$.

$$
\begin{array}{r}
4x^2 + 8x + 13 \\
x - 2\overline{\smash{)}4x^3 + 0x^2 - 3x + 6} \\
\underline{4x^3 - 8x^2} \\
8x^2 - 3x \\
\underline{8x^2 - 16x} \\
13x + 6 \\
\underline{13x - 26} \\
32
\end{array}
$$

$\leftarrow \frac{13x}{x} = 13$

$\leftarrow 13(x - 2)$

$\leftarrow$ Subtract.

We write the remainder, 32, as the numerator of the fraction $\dfrac{32}{x - 2}$. This is similar to when we convert an improper fraction to a mixed number. In summary,

$$
\frac{4x^3 - 3x + 6}{x - 2} = 4x^2 + 8x + 13 + \frac{32}{x - 2}.
$$

Check by multiplying $x - 2$ and $4x^2 + 8x + 13$ and adding 32 to the result. You should get $4x^3 - 3x + 6$. (Remember the + sign when adding the remainder to a quotient.)

If the polynomial divisor has a missing term, add a term as a placeholder.

EXAMPLE 5

Divide $6p^4 + 14p^3 - 25p^2 - 19p + 17$ by $2p^2 - 3$.

The polynomial $2p^2 - 3$ has a missing term. Write it as $2p^2 + 0p - 3$ and divide as usual.

$$3p^2 + 7p - 8$$

$$2p^2 + 0p - 3\overline{)6p^4 + 14p^3 - 25p^2 - 19p + 17}$$

$$\underline{6p^4 + 0p^3 - 9p^2}$$

$$14p^3 - 16p^2 - 19p$$

$$\underline{14p^3 + 0p^2 - 21p}$$

$$-16p^2 + 2p + 17$$

$$\underline{-16p^2 + 0p + 24}$$

$$2p - 7$$

Since the degree of the remainder, $2p - 7$, is less than the degree of $2p^2 - 3$, the division process is now finished. The result is written

$$3p^2 + 7p - 8 + \frac{2p - 7}{2p^2 - 3}.$$

Remember the following steps when dividing a polynomial by a polynomial of two or more terms.

1. Be sure the terms in both polynomials are in descending powers.
2. Write any missing terms with a 0 placeholder.

Self-Check 3

Divide.

1. $\dfrac{9x^4 + 8x^2 - 12x + 1}{3x - 1}$

2. $\dfrac{2d^4 - 3d^3 - 2d^2 - 7d + 5}{d^2 + 1}$

Self-Check Answers

1.1 $7x^2 - 2x + 3$

1.2 $8d^3 + 4d^2 - \dfrac{2}{d^2}$

1.3 $3f^2 - \dfrac{25f}{6} - \dfrac{3}{2f}$

1.4 $2ab + 1 - \dfrac{7}{5ab}$

2.1 $3x + 5$

2.2 $12x^2 + 25x - 7$

3.1 $3x^3 + x^2 + 3x - 3 - \dfrac{2}{3x - 1}$

3.2 $2d^2 - 3d - 4 + \dfrac{-4d + 9}{d^2 + 1}$

3.6 EXERCISES

Complete each statement with the correct word or words in exercises 1-4.

1. We find the quotient of two monomials by using the _____ rule for _____.

2. To divide a polynomial by a monomial, divide _____ of the polynomial by the _____.

3. When dividing polynomials that are not monomials, first write them in _____.

4. If a polynomial in a division problem has a missing term, insert a term with _____ as a placeholder.

Divide in exercises 5-12.

5. $\dfrac{25y^2 + 15y - 20}{5y}$

6. $\dfrac{16r^2 - 20r - 4}{4r}$

7. $\dfrac{12x^3 - 20x^2 + 8x}{8x^2}$

8. $\dfrac{49p^3 + 14p^2 + 12p}{14p^3}$

9. $\dfrac{21x^3y^2 - 12x^2y + 8xy^2}{14xy^2}$

10. $\dfrac{12m^4n^3 - 30m^3n^2 - 4mn^2}{20m^2n^2}$

11. $\dfrac{10x^3y^2z^3 + 2x^2y^3z + 5xyz^2}{5x^2y^2z^2}$

12. $\dfrac{12r^4s^3t + 3rs^3t + 6rs^3t^2}{6r^2st^3}$

Complete the division in exercises 13-14.

13.
$$
\begin{array}{r}
x^2 \\
2x - 1 \overline{)\, 2x^3 - 7x^2 + 11x - 4} \\
\underline{2x^3 - x^2 } \\
-6x^2
\end{array}
$$

14.
$$
\begin{array}{r}
p^2 \\
3p + 4 \overline{)\, 3p^3 + 10p^2 - 19p - 36} \\
\underline{3p^3 + 4p^2 } \\
6p^2
\end{array}
$$

Divide in exercises 15-30.

15. $\dfrac{y^2 - 2y - 8}{y + 2}$

16. $\dfrac{r^2 + 5r + 6}{r + 3}$

17. $\dfrac{2m^2 - 13m - 24}{m - 8}$

18. $\dfrac{7h^2 + 47h + 30}{7h + 5}$

19. $\left(6x^3 - 10x^2 - 3x + 5\right) \div \left(3x - 5\right)$

20. $\left(6z^3 + 5z^2 - 15z - 8\right) \div \left(2z + 1\right)$

21. $\left(3t^3 + 22t^2 - 32\right) \div \left(3t + 4\right)$

22. $\left(2z^3 - 17z + 3\right) \div \left(z + 3\right)$

23. $\dfrac{8y^3 + 8y^2 + 4y + 7}{2y + 3}$

24. $\dfrac{15r^3 - 28r^2 - 72r - 27}{5r + 4}$

25. $\dfrac{6b^4 - 10b^3 + 5b^2 - 15b - 6}{2b^2 + 3}$

26. $\dfrac{6a^4 + 18a^3 - 5a^2 + 30a - 25}{3a^2 + 5}$

27. $\dfrac{2p^4 - p^3 + p^2 + 7p + 1}{p^2 - 2p + 3}$

28. $\dfrac{c^4 - 7c^3 + 5c^2 - 65c + 34}{c^2 + c + 9}$

29. $\dfrac{p^3 - 8}{p - 2}$

30. $\dfrac{27g^3 + 8}{3g + 2}$

CH 3 | Summary

KEY TERMS

3.1 product rule for exponents   power rules for exponents
        quotient rule for exponents   scientific notation

3.2 term   binomial
        numerical coefficient (coefficient)   trinomial
        algebraic expression   degree of a term
        polynomial   degree of a polynomial
        polynomial in x descending powers   monomial

3.3 product of the sum and   square of a binomial
        difference of two terms

3.4 factoring   difference of two squares
        greatest common factor (GCF)   perfect square trinomial
        factoring out the greatest   difference of two cubes
            common factor   sum of two cubes
        factoring by grouping

3.5 factoring trinomials   prime polynomial

3.6 dividing a polynomial by   missing terms
            a monomial   quotient
        remainder

CH 3 | Quick Review

3.1 INTEGER EXPONENTS AND SCIENTIFIC NOTATION
Definitions and Rules for Exponents

Product Rule: $a^m \cdot a^n = a^{m+n}$

Quotient Rule: $\dfrac{a^m}{a^n} = a^{m-n}$

Negative Exponent: $a^{-n} = \dfrac{1}{a^n}$ $\dfrac{a^{-n}}{b^{-m}} = \dfrac{b^m}{a^n}$

Zero Exponent: $a^0 = 1, \ a \neq 0$

Power Rules: $\left(a^m\right)^n = a^{mn}$ $(ab)^m = a^m b^m$

$\left(\dfrac{a}{b}\right)^n = \dfrac{a^n}{b^n}$ $a^{-n} = \left(\dfrac{1}{a}\right)^n$

$\left(\dfrac{a}{b}\right)^{-n} = \left(\dfrac{b}{a}\right)^n$

Scientific Notation
A number is in scientific notation when it is written as a product of a number between 1 and 10 (inclusive of 1) and an integer power of 10.

3.2 ADDITION AND SUBTRACTION OF POLYNOMIALS
Add or subtract polynomials by combining like terms.

3.3 MULTIPLICATION OF POLYNOMIALS

To multiply two polynomials, multiply each term of one by each term of the other.

Special Products

$$(x - y)(x + y) = x^2 - y^2, \; (x + y)^2 = x^2 + 2xy + y^2, \; \text{and} \; (x - y)^2 = x^2 - 2xy + y^2$$

To multiply two binomials in general, use the FOIL method. Multiply the First terms, the Outside terms, the Inside terms, and the Last terms.

3.4 INTRODUCTION TO FACTORING

Factoring Out the Greatest Common Factor
The product of the largest common numerical factor and the variable of lowest degree common to every term in a polynomial is the greatest common factor of the terms of the polynomial.

Factoring by Grouping
Group the terms so that each group has a common factor. Factor out the common factor in each group. If the groups now have a common factor, factor it out. If not, try a different grouping.

Difference of Two Squares

$$x^2 - y^2 = (x - y)(x + y)$$

Perfect Square Trinomials

$$x^2 + 2xy + y^2 = (x + y)^2$$
$$x^2 - 2xy + y^2 = (x - y)^2$$

Difference of Two Cubes

$$x^3 - y^3 = (x - y)(x^2 + xy + y^2)$$

Sum of Two Cubes

$$x^3 + y^3 = (x + y)(x^2 - xy + y^2)$$

3.5 FACTORING TRINOMIALS

To factor a trinomial, choose factors of the first term and factors of the last term. Then, place them in a pair of parentheses of the form:( )( ). Try various combinations of the factors until the correct middle term of the trinomial is found.

3.6 DIVISION OF POLYNOMIALS

Dividing by a Monomial
To divide a polynomial by a monomial, divide each term in the polynomial by the monomial, and then write each fraction in lowest terms.

Dividing by a Polynomial
Use the "long division" process. Use the following steps when dividing a polynomial by a polynomial of two or more terms.

Step 1   Be sure the terms in both polynomials are in descending powers.
Step 2   Write any missing terms with a 0 placeholder.
Step 3   Divide the first term of the dividend by the first term of the divisor. This quotient is the first term of the resulting quotient.
Step 4   Multiply the result found in step 2 by the divisor and be sure to place terms of this product under the like terms in the dividend.
Step 5   Subtract the product found in step 4 from the dividend.
Step 6   Bring down the next term in the dividend and continue the process as in division of numbers.
Step 7   This process should continue until either the divisor "goes evenly" into the dividend or the degree of the remainder is less than the degree of the divisor.
Step 8   Write the result and don't forget to include any remainder.

CH 3 | Review Exercises

Use the product rule and/or the quotient rule to simplify exercises 1-3. Write the answers with only positive exponents. Assume that all variables represent nonzero real numbers.

1. $\left(-4x^2y^3\right)\left(2x^{-3}y^7\right)$ **2.** $\dfrac{8b^{-3}c^5}{2b^2c^4}$ **3.** $\dfrac{\left(2h^{-2}k^5\right)\left(8h^5k^{-3}\right)}{4h^{-4}k^5}$

4. Explain the difference between the expressions $(-5)^0$ and -5^0.

Evaluate exercises 5-13.

5. 2^4 **6.** $\left(\dfrac{1}{3}\right)^3$ **7.** $(-2)^3$

8. $\dfrac{4}{(-5)^{-2}}$ **9.** $\left(\dfrac{3}{4}\right)^{-2}$ **10.** $\left(\dfrac{2}{3}\right)^{-3}$

11. $2^{-1} + 4^{-1}$ **12.** $(2+4)^{-1}$ **13.** $-6^0 + 6^0$

14. Give an example to show that $(2a)^{-3}$ is not equal to $\dfrac{2}{a^{-3}}$ in general by choosing a specific value for a. (Avoid using $a = 0$.)

Simplify exercises 15-22. Write answers with only positive exponents. Assume that all variables represent positive real numbers.

15. $\left(2^{-3}\right)^2$ **16.** $\left(x^{-3}\right)^{-4}$ **17.** $\left(xy^{-2}\right)^{-3}$

18. $\left(z^{-4}\right)^2 z^{-8}$ **19.** $\left(6p^{-5}\right)^2\left(p^{-4}\right)^{-8}$ **20.** $\dfrac{(2w)^3 w^5}{w^0 w^{-5}}\left(3w^{-4}\right)^{-3}$

21. $\left(\dfrac{3k^{-3}}{k^{-4}}\right)^{-1}\left(\dfrac{k^{-5}}{12}\right)$ **22.** $\left(\dfrac{2j^4}{j^{-5}}\right)^{-2}\left(\dfrac{j^{-2}}{6}\right)^2$

23. Is $\left(\dfrac{a}{b}\right)^{-1} = \dfrac{a^{-1}}{b^{-1}}$ for all $a, b \neq 0$? If not, explain.

24. Is $(ab)^{-1} = ab^{-1}$ for all $a, b \neq 0$? If not, explain.

25. Give an example to show that $\left(x^2 + y^2\right)^2 \neq x^4 + y^4$ by choosing specific values for x and y.

Write exercises 26-28 in scientific notation.

26. 14,200   **27.** 0.000000175   **28.** 0.0125

Write exercises 29-30 without scientific notation.

29. 2.5×10^5

30. 3.18×10^{-4}

31. Light travels 1.86×10^5 miles per second. Write this figure without using scientific notation.

Use scientific notation to compute exercises 32-35. Give answers in both scientific notation and standard form.

32. $\dfrac{12 \times 10^9}{6 \times 10^{12}}$

33. $\dfrac{8 \times 10^{-6}}{5 \times 10^{-9}}$

34. $\dfrac{0.00000000000128}{0.0000002}$

35. $\dfrac{9,000,000,000 \times 0.00000000012}{0.00000036}$

36. A light-year is the distance that light travels in one year. Find the number of miles in a light year. (Assume there are 365 days per year and refer to exercise 31)

Give the (numerical) coefficient for each term in exercises 37-39.

37. $17x^2$

38. $-y^3$

39. $14x^2y^3$

*For the polynomial in exercises 40-43, **(a)** write in descending powers, **(b)** identify as monomial, binomial, trinomial, or none of these, and **(c)** give the degree.*

40. $5k^5 - 3k^7 - 1$

41. $2a^5 - 9a^7$

42. $3r^2 + 6r^9 + 5r^3 - r + 2$

43. l^9

44. Give an example of a polynomial in the variable x such that it has degree 6, is lacking a second-degree term, and is in descending powers of the variable.

Add or subtract as indicated in exercises 45-48.

45. Add
$$\begin{array}{r} 2x^2 - 3x - 1 \\ -3x^2 + 2x - 9 \end{array}$$

46. Subtract
$$\begin{array}{r} -y^4 \quad\;\; + 4y^2 + 2y - 1 \\ 2y^3 + 7y^2 - 8y - 2 \end{array}$$

47. $\left(5r^3 - 2r + 12\right) - \left(7r^3 + 9r^2 - r\right)$

48. $\left(8y^2 + 7y - 6\right) + \left(-5y^2 - 3y + 9\right)$

49. Find the perimeter of the triangle.

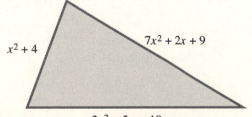

$x^2 + 4$

$7x^2 + 2x + 9$

$2x^2 + 5x + 10$

Find each product in exercises 50-60.

50. $-5k\left(3k^2 + 9\right)$

51. $(2x - 7)(3x + 1)$

52. $(3y - 5)(y - 8)$

53. $(7s - 6t)(2s + 5t)$

54. $\left(3p^2 - 4\right)\left(4p^2 + 5\right)$

55. $\left(4x^2 - 2x - 5\right)(2x + 1)$

56. $\left(2z^3 + 5z^2 - 3z + 1\right)(3z - 2)$

57. $\left(2k^2 - 5\right)\left(2k^2 + 5\right)$

58. $\left(g + \dfrac{2}{3}\right)\left(g - \dfrac{2}{3}\right)$

59. $(3c + 1)^2$

60. $(2d - 7)^2$

Factor out the greatest common factor in exercises 61-66.

61. $2x^2 - 4x$

62. $15k^5 - 20k^7$

63. $6a^2b - 9a^3b^2 + 12ab^2$

64. $12st^2 - 24s^4t^5 - 8s^2t^3$

65. $(a + 3)(a - 1) + (a + 3)(2a - 5)$

66. $(x + 4)(x - 9) - (x + 4)(7x - 6)$

Factor by grouping in exercises 67-70.

67. $5am - 5bm - an + bn$

68. $x^2 + 4z + 4x + xz$

69. $8p + nr + pn + 8r$

70. $3m + 9 - mn - 3n$

Factor completely in exercises 71-78.

71. $6x^2 + 23x + 20$

72. $9p^2 - 9p - 4$

73. $14y^2 + 47y - 7$

74. $4b^2 - 24b + 27$

75. $9x^2 - 6xy - 8y^2$

76. $3a^2 - 11ab + 10b^2$

77. $42x^3 - 82x^2 + 20x$

78. $9r^3 - 21r^2 + 6r$

79. When asked to factor $x^2y^2 - 4x^2 + 6y^2 - 24$, a student gave the following incorrect answer: $x^2\left(y^2 - 4\right) + 6\left(y^2 - 4\right)$. Why is this answer incorrect? What is the correct answer?

80. If the area of this rectangle is represented by $12s^2 - 7s + 1$, what is the width in terms of s?

$3s - 1$

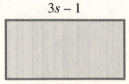

Factor completely in exercises 81-90.

81. $25x^2 - 16$　　　　**82.** $100y^2 - 49z^2$　　　　**83.** $49y^2 + 70y + 25$

84. $25d^2 - 30d + 9$　　**85.** $x^3 + 125$　　　　**86.** $64s^3 - 1$

87. $x^6 + 8$　　　　　　**88.** $x^2 + 4x + 4 - 25y^2$　**89.** $(s + t)^3 - (s - t)^3$

90. $(a + 1)^3 + (a - 1)^3$

Divide in exercises 91-94.

91. $\dfrac{8x^6 - 6x^5 - 4x^2}{4x^3}$　　　　　　**92.** $\dfrac{6x^3 - 19x^2 + 5}{2x - 5}$

93. $\dfrac{12x^5 - 38x^4 + 42x^3 - 123x^2 + 35}{2x^2 + 7}$　　**94.** $\dfrac{12x^4 - 7}{2x^2 - 5}$

CH 3 ‖ Test

1. Match the expression in Column I with its equivalent expression from Column II. Choices may be used once, more than once, or not at all.

	I		II
(a)	8^{-2}	A.	$\dfrac{1}{11}$
(b)	8^0	B.	-1
(c)	-8^0	C.	1
(d)	$(-8)^0$	D.	-64
(e)	-8^2	E.	$\dfrac{1}{64}$
(f)	$8^{-1} + 3^{-1}$	F.	$\dfrac{11}{24}$
(g)	$(8 + 3)^{-1}$	G.	$\dfrac{3}{8}$
(h)	$\dfrac{8^{-1}}{3^{-1}}$	H.	0
(i)	$(-8)^{-2}$	I.	none of these

Simplify exercises 2-5. Write answers with only positive exponents. Assume that all variables represent positive real numbers.

2. $\left(3x^{-3}y^4\right)^{-2}\left(2x^{-3}y^5\right)$　　　　**3.** $\dfrac{\left(12h^4k^{-3}\right)\left(3h^0k^{-3}\right)}{9h^{-6}k^{-5}}$

4. $\left(\dfrac{3j^5}{j^{-7}k^2}\right)^{-3}\left(\dfrac{j^3k^7}{4}\right)^{-2}$　　　**5.** $\left(-2x^2y^0z^{-4}\right)^{-3}\left(3x^2\right)^0$

6.   (a)   Write 1.0134×10^6 without scientific notation.

   (b)   Use scientific notation to simplify $\dfrac{120{,}000{,}000 \times 0.0000009}{0.000024 \times 0.003}$. Write the answer in
          both standard and scientific notation.

Perform the indicated operations in exercises 7-11.

7. $\left(8x^3 - 8x + 1\right) - \left(5x^3 - 7x^2 - 3x\right) + \left(x^2 - 1\right)$

8. $(5x - 9)(3x + 7)$ **9.** $\left(5x^2 - 4x - 3\right)(2x - 1)$

10. $(3m - 4e)(3m + 4e)$ **11.** $(5a - 9)^2$

Factor exercises 12-18.

12. $121a^3 - 132a^4$ **13.** $3p^2 - 17p + 10$

14. $6x^2 + 13xy - 33y^2$ **15.** $9x^2 - 24x + 16$

16. $9j^2 - 49k^2$ **17.** $x^3 + 216$

18. $84x^3y - 164x^2y + 40xy$

Divide in exercises 19-20.

19. $\dfrac{12x^5y^2 - 24x^4y^3 - 36x^3}{8x^3y^2}$ **20.** $\dfrac{9x^3 - 21x^2 + 5}{3x - 1}$

CHAPTER 4: RATIONAL EXPRESSIONS

4.1 | Fractions

As preparation for the study of rational expressions, this section begins with a brief review of fractions.

The parts of a fraction are named as follows.

$$\text{Fraction bar} \rightarrow \frac{8}{9} \quad \left(\frac{a}{b} = a \div b \right) \quad \begin{matrix} \leftarrow \text{ Numerator} \\ \leftarrow \text{ Denominator} \end{matrix}$$

As we will see later, the fraction bar represents division and also serves as a grouping symbol.

Write Fractions in Lowest Terms: The number 12 is **factored** by writing it as the product of two or more numbers. For example, 12 can be factored in several ways, as $6 \cdot 2$, $4 \cdot 3$, $12 \cdot 1$, or $2 \cdot 2 \cdot 3$.

A natural number (except 1) is **prime** if it has only itself and 1 as factors. "Factors" are understood here to mean natural number factors. For example, we can write 35 in factored form by writing it as a product of prime factors: $35 = 5 \cdot 7$. Similarly $24 = 2 \cdot 2 \cdot 2 \cdot 3$, where all factors are prime.

We use prime numbers to write fractions in *lowest terms*. A fraction is in **lowest terms** when the numerator and denominator have no factors in common (other than 1). By the **basic principle of fractions**, if the numerator and denominator of a fraction are multiplied or divided by the *same* nonzero number, the value of the fraction is unchanged. To write a fraction in lowest terms, use these steps.

Procedure: Writing a Fraction in Lowest Terms

Step 1 Write the numerator and the denominator as the product of prime factors.

Step 2 Divide the numerator and the denominator by the **greatest common factor**, the product of all factors common to both.

EXAMPLE 1
Write the fraction in lowest terms.

(a) $\dfrac{5}{15} = \dfrac{5}{3 \cdot 5} = \dfrac{1}{3 \cdot 1} = \dfrac{1}{3}$

Since 5 is the greatest common factor of 5 and 15, dividing both numerator and denominator by 5 gives the fraction in lowest terms.

(b) $\dfrac{24}{30} = \dfrac{2 \cdot 2 \cdot 2 \cdot 3}{2 \cdot 3 \cdot 5} = \dfrac{2 \cdot 2 \cdot 1 \cdot 1}{1 \cdot 1 \cdot 5} = \dfrac{4}{5}$

The factored form shows that 2 and 3 are the common factors of both 24 and 30. Dividing both 24 and 30 by $2 \cdot 3 = 6$ gives $\frac{24}{30}$ in lowest terms, as $\frac{4}{5}$.

We can simplify this process by finding the greatest common factor in the numerator and denominator by inspection. For instance, in Example 1(b), we can use 6 rather than $2 \cdot 3$.

$$\frac{24}{30} = \frac{4 \cdot 6}{5 \cdot 6} = \frac{4 \cdot 1}{5 \cdot 1} = \frac{4}{5}$$

Errors may occur when writing fractions in lowest terms if the factor 1 is not included. To see this, refer to Example 1(a). In the equation,

$$\frac{5}{3 \cdot 5} = \frac{?}{3},$$

if 1 is not written in the numerator when dividing common factors, you may make an error. The ? should be replaced by 1.

Self-Check 1

Write fraction in lowest terms.

1. $\dfrac{48}{80}$ 2. $\dfrac{18}{27}$ 3. $\dfrac{20}{40}$ 4. $\dfrac{40}{20}$

Multiply and Divide Fractions: The basic operations on whole numbers, addition, subtraction, multiplication, and division, also apply to fractions. We multiply two fractions by first multiplying their numerators and then multiplying their denominators. This rule is written in symbols as follows.

Rule: Multiplying Fractions

If $\dfrac{a}{b}$ and $\dfrac{c}{d}$ are fractions, then $\dfrac{a}{b} \cdot \dfrac{c}{d} = \dfrac{ac}{bd}$.

EXAMPLE 2

Find the product of $\frac{5}{6}$ and $\frac{3}{25}$, and write it in lowest terms.

First, multiply $\frac{5}{6}$ and $\frac{3}{25}$.

$$\frac{5}{6} \cdot \frac{3}{25} = \frac{5 \cdot 3}{6 \cdot 25} \quad \text{Multiply numerators; multiply denominators.}$$

It is easiest to write a fraction in lowest terms while the product is in factored form. Factor the 6 and 25 and then divide out common factors in the numerator and denominator.

$$\frac{5 \cdot 3}{6 \cdot 25} = \frac{1 \cdot 5 \cdot 3}{2 \cdot 3 \cdot 5 \cdot 5} \quad \text{Factor. Introduce a factor of 1.}$$

$$= \frac{1}{2 \cdot 5} \quad \text{3 and 5 are common factors.}$$

$$= \frac{1}{10} \quad \text{Lowest terms}$$

Two fractions are **reciprocals** of each other if their product is 1. For example, $\frac{2}{5}$ and $\frac{5}{2}$ are reciprocals since

$$\frac{2}{5} \cdot \frac{5}{2} = \frac{10}{10} = 1.$$

Also, $-\frac{8}{9}$ and $-\frac{9}{8}$ are reciprocals of each other. We use the reciprocal to divide fractions. To *divide* two fractions, multiply the first fraction by the reciprocal of the second fraction.

Rule: Dividing Fractions

$$\text{If } \frac{a}{b} \text{ and } \frac{c}{d} \text{ are fractions, then } \frac{a}{b} \div \frac{c}{d} = \frac{a}{b} \cdot \frac{d}{c}.$$

(To divide by a fraction, multiply by its reciprocal)

The answer to a division problem is called a **quotient**. For example, the quotient of 30 and 6 is 5, since $30 \div 6 = 5$.

EXAMPLE 3

Find the following quotients, and write them in lowest terms.

(a) $\dfrac{5}{7} \div \dfrac{14}{9} = \dfrac{5}{7} \cdot \dfrac{9}{14} = \dfrac{5 \cdot 9}{7 \cdot 14} = \dfrac{45}{98}$ Multiply by the reciprocal of $\frac{14}{9}$.

(b) $\dfrac{9}{14} \div \dfrac{5}{7} = \dfrac{9}{14} \cdot \dfrac{7}{5} = \dfrac{9 \cdot 7}{14 \cdot 5} = \dfrac{9 \cdot 7}{7 \cdot 2 \cdot 5} = \dfrac{9}{10}$

Self-Check 2

Find the product of the following and write in lowest terms.

1. $\dfrac{1}{2} \cdot \dfrac{7}{8}$ **2.** $\dfrac{4}{5} \cdot \dfrac{25}{32}$

Find the quotient of the following and write in lowest terms.

3. $\dfrac{4}{5} \div \dfrac{4}{3}$ **4.** $\dfrac{5}{7} \div 20$

Add and Subtract Fractions: To find the **sum** of two fractions having the same denominator, add the numerators and keep the same denominator.

Rule: Adding Fractions

$$\text{If } \frac{a}{b} \text{ and } \frac{c}{b} \text{ are fractions, then } \frac{a}{b} + \frac{c}{b} = \frac{a+c}{b}.$$

EXAMPLE 4

(a) $\dfrac{2}{9} + \dfrac{5}{9} = \dfrac{2+5}{9} = \dfrac{7}{9}$ Add numerators and keep the same denominator.

(b) $\dfrac{1}{8} + \dfrac{3}{8} = \dfrac{1+3}{8} = \dfrac{4}{8} = \dfrac{1}{2}$

If the fractions to be added do not have the same denominators, the procedure above can still be used, but only *after* the fractions are rewritten with a common denominator. For example, to rewrite $\frac{2}{3}$ as a fraction with a denominator of 15,

$$\frac{2}{3} = \frac{?}{15},$$

find the number that can be multiplied by 3 to give 15. Since $3 \cdot 5 = 15$, use the number 5. By the basic principle, we can multiply the numerator and the denominator by 5.

$$\frac{2}{3} = \frac{2 \cdot 5}{3 \cdot 5} = \frac{10}{15}$$

Procedure: Finding the Least Common Denominator

To add (or subtract) fractions with different denominators, find the **least common denominator (LCD)** as follows.

Step 1   Factor both denominators.

Step 2   For the LCD, use every factor that appears in any factored form. If a factor is repeated, use the largest number of repeats in the LCD.

The next example shows this procedure used when the fractions have different denominators.

EXAMPLE 5

Add the fractions, $\dfrac{5}{12}$ and $\dfrac{2}{15}$.

To find the least common denominator, first factor both denominators.

$$12 = 2 \cdot 2 \cdot 3 \quad \text{and} \quad 15 = 3 \cdot 5$$

Since 2, 3 and 5 appear as factors, and 2 is a factor of 12 twice, the LCD is

$$2 \cdot 2 \cdot 3 \cdot 5 \quad \text{or} \quad 60.$$

Write each fraction with 60 as denominator.

$$\frac{5}{12} = \frac{5 \cdot 5}{12 \cdot 5} = \frac{25}{60} \quad \text{and} \quad \frac{2}{15} = \frac{2 \cdot 4}{15 \cdot 4} = \frac{8}{60}$$

Now add the two equivalent fractions.

$$\frac{25}{60} + \frac{8}{60} = \frac{25 + 8}{60} = \frac{33}{60} = \frac{11}{20}$$

The difference between two numbers is found by subtraction. For example, $10 - 8 = 2$ so the difference between 10 and 8 is 2. Subtraction of fractions is similar to addition. Just subtract the numerators instead of adding them; again, keep the same denominator.

Rule: Subtracting Fractions

If $\dfrac{a}{b}$ and $\dfrac{c}{b}$ are fractions, then $\dfrac{a}{b} - \dfrac{c}{b} = \dfrac{a - c}{b}$.

EXAMPLE 6

Subtract. Write the differences in lowest terms.

(a) $\dfrac{13}{15} - \dfrac{1}{15} = \dfrac{13-1}{15}$ Subtract numerators; keep the same denominator.

$= \dfrac{12}{15} = \dfrac{4}{5}$ Lowest terms

(b) $\dfrac{17}{21} - \dfrac{3}{28}$

Here, $21 = 3 \cdot 7$ and $28 = 2 \cdot 2 \cdot 7$, so the LCD is $2 \cdot 2 \cdot 3 \cdot 7 = 84$.

$$\frac{17}{21} - \frac{3}{28} = \frac{17 \cdot 2 \cdot 2}{2 \cdot 2 \cdot 3 \cdot 7} - \frac{3 \cdot 3}{2 \cdot 2 \cdot 3 \cdot 7} = \frac{68}{84} - \frac{9}{84} = \frac{59}{84}$$

(c) $\dfrac{17}{27} - \dfrac{7}{20}$

Since $27 = 3 \cdot 3 \cdot 3$ and $20 = 2 \cdot 2 \cdot 5$, there are no common factors, and the LCD is $27 \cdot 20 = 540$.

$$\frac{17}{27} - \frac{7}{20} = \frac{17 \cdot 20}{27 \cdot 20} - \frac{7 \cdot 27}{27 \cdot 20}$$ Get a common denominator.

$$= \frac{340}{540} - \frac{189}{540}$$

$$= \frac{151}{540}$$ Subtract.

Self-Check 3

Find the sum of the following and write in lowest terms.

1. $\dfrac{7}{12} + \dfrac{3}{12}$ 2. $\dfrac{7}{20} + \dfrac{13}{36}$

Find the difference of the following and write in lowest terms.

3. $\dfrac{17}{21} - \dfrac{5}{21}$ 4. $\dfrac{29}{30} - \dfrac{7}{24}$

Self-Check Answers

1.1 $\dfrac{3}{5}$ 1.2 $\dfrac{2}{3}$ 1.3 $\dfrac{1}{2}$ 1.4 2

2.1 $\dfrac{7}{16}$ 2.2 $\dfrac{5}{8}$ 2.3 $\dfrac{3}{5}$ 2.4 $\dfrac{1}{28}$

3.1 $\dfrac{5}{6}$ 3.2 $\dfrac{32}{45}$ 3.3 $\dfrac{4}{7}$ 3.4 $\dfrac{27}{40}$

4.1 EXERCISES

Decide whether each statement is true or false in exercises 1-4. If it is false, say why.

1. In the fraction $\dfrac{4}{9}$, 4 is the numerator and 9 is the denominator.

2. The fraction, $\dfrac{14}{91}$, is in lowest terms.

3. The reciprocal of $\dfrac{15}{3}$ is $\dfrac{5}{1}$.

4. The difference between $\dfrac{5}{6}$ and $\dfrac{1}{2}$ is $\dfrac{4}{4} = 1$.

Write each fraction in lowest terms in exercises 5-12.

5. $\dfrac{10}{20}$ 6. $\dfrac{6}{18}$ 7. $\dfrac{6}{16}$ 8. $\dfrac{24}{40}$

9. $\dfrac{24}{64}$ 10. $\dfrac{168}{204}$ 11. $\dfrac{143}{165}$ 12. $\dfrac{96}{184}$

13. One of the following is the correct way to write $\dfrac{16}{24}$ in lowest terms. Which one is it?

   (a) $\dfrac{16}{24} = \dfrac{8+8}{8+16} = \dfrac{8}{16} = \dfrac{1}{2}$ (b) $\dfrac{16}{24} = \dfrac{4 \cdot 4}{4 \cdot 6} = \dfrac{4}{6}$

   (c) $\dfrac{16}{24} = \dfrac{8 \cdot 2}{8 \cdot 3} = \dfrac{2}{3}$ (d) $\dfrac{16}{24} = \dfrac{14+2}{21+3} = \dfrac{2}{3}$

14. For the fractions $\dfrac{p}{q}$ and $\dfrac{r}{s}$, which one of the following can serve as a common denominator?

   (a) $q \cdot s$ (b) $q + s$ (c) $p \cdot r$ (d) $p + r$

Find each product or quotient, and write in lowest terms in exercises 15-30.

15. $\dfrac{2}{3} \cdot \dfrac{4}{5}$ 16. $\dfrac{7}{8} \cdot \dfrac{9}{11}$ 17. $\dfrac{1}{8} \cdot \dfrac{12}{7}$ 18. $\dfrac{12}{7} \cdot \dfrac{13}{6}$

19. $\dfrac{6}{5} \cdot \dfrac{10}{27}$ 20. $\dfrac{8}{3} \cdot \dfrac{27}{5}$ 21. $\dfrac{11}{12} \cdot \dfrac{81}{22}$ 22. $\dfrac{21}{16} \cdot \dfrac{64}{15}$

23. $\dfrac{1}{4} \div \dfrac{7}{8}$ 24. $\dfrac{3}{2} \div \dfrac{1}{10}$ 25. $\dfrac{1}{2} \div \dfrac{4}{5}$ 26. $\dfrac{3}{5} \div \dfrac{25}{9}$

27. $\dfrac{3}{4} \div \dfrac{9}{16}$ 28. $\dfrac{7}{8} \div \dfrac{49}{12}$ 29. $\dfrac{9}{4} \div 12$ 30. $\dfrac{5}{7} \div 15$

31. Write a summary explaining how to multiply and divide fractions. Give examples.

32. Write a summary explaining how to add and subtract fractions. Give examples.

Find each sum or difference, and write in lowest terms in exercises 33-44.

33. $\dfrac{7}{20} + \dfrac{3}{20}$ **34.** $\dfrac{1}{9} + \dfrac{2}{9}$ **35.** $\dfrac{3}{16} + \dfrac{5}{8}$ **36.** $\dfrac{2}{5} + \dfrac{7}{20}$

37. $\dfrac{15}{16} - \dfrac{1}{16}$ **38.** $\dfrac{19}{21} - \dfrac{4}{21}$ **39.** $\dfrac{23}{24} - \dfrac{1}{8}$ **40.** $\dfrac{13}{16} - \dfrac{1}{4}$

41. $\dfrac{10}{9} + \dfrac{7}{10} - \dfrac{5}{6}$ **42.** $\dfrac{11}{12} - \dfrac{3}{8} + \dfrac{1}{6}$ **43.** $\dfrac{3}{4} - \dfrac{1}{8} + \dfrac{5}{6}$ **44.** $\dfrac{14}{15} - \dfrac{1}{5} - \dfrac{1}{20}$

4.2 Rational Expressions: Multiplication and Division

Rational Expressions: In arithmetic, a rational number is the quotient of two integers, with the denominator not 0. In algebra, a **rational expression** or algebraic fraction is the quotient of two polynomials, again with the denominator not 0. For example,

$$\frac{x+3}{x-5}, \quad \frac{3y^2 - 7y + 2}{5y^2 - 8}, \quad \text{and} \quad b^9 \left(\text{or } \frac{b^9}{1} \right)$$

are all rational expressions. In other words, rational expressions are the elements of the set

$$\left\{ \frac{P}{Q} \,\middle|\, P, Q \text{ polynomials, with } Q \neq 0 \right\}$$

The denominator of a rational expression cannot equal zero. So, for $\dfrac{x+3}{x-5}$, x cannot equal 5.

Rational Expressions in Lowest Terms: In arithmetic, we write the fraction $\frac{8}{20}$ in lowest terms by dividing the numerator and denominator by 4 to get $\frac{2}{5}$. We write rational expressions in lowest terms in a similar way, using the **fundamental principle of rational numbers**.

Fundamental Principle of Rational Numbers

If $\frac{a}{b}$ is a rational number and if c is any nonzero real number, then

$$\frac{a}{b} = \frac{ac}{bc}.$$

(The numerator and denominator of a rational number may either be multiplied or divided by the same nonzero number without changing the value of the rational number.)

Since $\frac{c}{c}$ is equivalent to 1, the fundamental principle is based on the identity property of multiplication.

A rational expression is a quotient of two polynomials, and since the value of a polynomial is a real number for all values of the variables for which it is defined, any statement that applies to rational numbers will also apply to rational expressions. We use the following steps to write rational expressions in lowest terms.

Procedure: Writing in Lowest Terms

Step 1   **Factor.** Factor both numerator and denominator to find their greatest common factor (GCF).

Step 2   **Reduce.** Apply the fundamental principle.

EXAMPLE 1
Write each rational expression in lowest terms.

(a) $\dfrac{4z}{24} = \dfrac{z \cdot 4}{6 \cdot 4} = \dfrac{z}{6}$

Here, the GCF of the numerator and denominator is 4. We then applied the fundamental principle.

(b) $\dfrac{x+9}{18}$

This expression is in lowest terms. Because the numerator cannot be factored, the expression cannot be simplified further.

(c) $\dfrac{24x^5 y}{12x^3 y^2} = \dfrac{2x^2 \cdot 12x^3 y}{y \cdot 12x^3 y} = \dfrac{2x^2}{y}$

Factor out the GCF, $12x^3 y$. (Here 3 is the least exponent on x, and 1 the least exponent on y.)

(d) $\dfrac{a^2 - 3a - 18}{a^2 - 7a + 6}$

Start by factoring the numerator and denominator.

$$\frac{a^2 - 3a - 18}{a^2 - 7a + 6} = \frac{(a+3)(a-6)}{(a-1)(a-6)}$$

Divide the numerator and denominator by $a - 6$ to get

$$\frac{a^2 - 3a - 18}{a^2 - 7a + 6} = \frac{a+3}{a-1}.$$

(e) $\dfrac{s^2 - 25}{s^2 - 3s - 10} = \dfrac{(s+5)(s-5)}{(s+2)(s-5)} = \dfrac{s+5}{s+2}$

Be careful! When using the fundamental principle of rational numbers, only common factors may be divided. For example,

$$\frac{x-4}{4} \neq x \quad \text{and} \quad \frac{x-4}{4} \neq x - 1$$

because the 4 in $x - 4$ is not a factor of the numerator. Remember to *factor* before writing a fraction in lowest terms.

EXAMPLE 2

Write the rational expression, $\dfrac{x-5}{5-x}$, in lowest terms.

In this rational expression, the numerator and denominator are opposites. The given expression can be written in lowest terms by writing the denominator as $5 - x = -1(x - 5)$, giving

$$\frac{x-5}{5-x} = \frac{x-5}{-1(x-5)} = \frac{1}{-1} = -1.$$

Self-Check 1

Write the following rational expressions in lowest terms.

1. $\dfrac{3x^2 y^4 z^7}{12xy^6 z^8}$
2. $\dfrac{x^2 - 2x - 8}{x^2 - 6x + 8}$
3. $\dfrac{2c^2 - 12c + 10}{4c - 4}$
4. $\dfrac{x^2 - 16}{4 - x}$

Multiply Rational Expressions: To multiply rational expressions, follow these steps. (In practice, we usually simplify before multiplying.)

Procedure: Multiplying Rational Expressions

Step 1   **Factor.** Factor all numerators and denominators as completely as possible.

Step 2   **Reduce.** Apply the fundamental principle.

Step 3   **Multiply.** Multiply remaining factors in the numerator and remaining factors in the denominator.

Step 4   **Check.** Check to be sure the product is in lowest terms..

EXAMPLE 3

Multiply.

(a) $\dfrac{5x^4}{2} \cdot \dfrac{8}{x^5} = \dfrac{5x^4 \cdot 8}{2 \cdot x^5} = \dfrac{40x^4}{2x^5} = \dfrac{20 \cdot 2x^4}{x \cdot 2x^4} = \dfrac{20}{x}$

Apply the fundamental principle using $2x^4$ to write the product in lowest terms. Notice that common factors in the numerator and denominator can be divided out *before* multiplying the numerator factors and the denominator factors as follows.

$$\frac{5x^4}{2} \cdot \frac{8}{x^5} = \frac{5 \cdot x^4}{2} \cdot \frac{4 \cdot 2}{x \cdot x^4} = \frac{20}{x}$$

This is the most efficient way to multiply fractions.

(b) $\dfrac{2c-2}{c} \cdot \dfrac{7c^2}{10c-10}$

First, factor where possible.

$$\frac{2c-2}{c} \cdot \frac{7c^2}{10c-10} = \frac{2(c-1)}{c} \cdot \frac{7c \cdot c}{5 \cdot 2(c-1)} \qquad \text{Factor.}$$

$$= \frac{1}{1} \cdot \frac{7c}{5} \qquad\qquad\qquad \text{Lowest terms}$$

$$= \frac{7c}{5} \qquad\qquad\qquad\qquad \text{Multiply.}$$

(c) $\dfrac{k^2 - 6k - 16}{k^2 - 16} \cdot \dfrac{k^2 - 4k}{k^2 + 3k + 2} = \dfrac{(k - 8)(k + 2)}{(k - 4)(k + 4)} \cdot \dfrac{k(k - 4)}{(k + 1)(k + 2)}$

$$= \dfrac{k(k - 8)}{(k + 4)(k + 1)}$$

(d) $(q - 8) \cdot \dfrac{7}{14q - 112} = \dfrac{q - 8}{1} \cdot \dfrac{7}{14q - 112}$ Write $q - 8$ as $\frac{q-8}{1}$.

$$= \dfrac{q - 8}{1} \cdot \dfrac{7}{14(q - 8)}$$ Factor.

$$= \dfrac{1}{2}$$

(e) $\dfrac{x^2 + 5x}{x^2 - 25} \cdot \dfrac{x^2 - 1}{x^3 + x^2} = \dfrac{x(x + 5)}{(x - 5)(x + 5)} \cdot \dfrac{(x - 1)(x + 1)}{x^2(x + 1)}$ Factor where possible.

$$= \dfrac{x(x + 5)(x - 1)(x + 1)}{x^2(x - 5)(x + 5)(x + 1)}$$ Multiply.

$$= \dfrac{x - 1}{x(x - 5)}$$ Lowest terms

Self-Check 2

Multiply.

1. $\dfrac{5x^2 y}{14} \cdot \dfrac{7y^2}{10x^3}$

2. $\dfrac{4a - 4}{a^2} \cdot \dfrac{3a}{16a - 16}$

3. $\dfrac{z^2 - 3z + 2}{z^2 - 2z + 1} \cdot \dfrac{z^2 - z}{2z^4}$

4. $(5x - 5) \cdot \dfrac{4x}{1 - x}$

Reciprocals of Rational Expressions: Recall that rational numbers $\frac{a}{b}$ and $\frac{c}{d}$ are reciprocals of each other if they have a product of 1. The **reciprocal** of a rational expression can be defined in the same way: Two rational expressions are reciprocals of each other if they have a product of 1. Recall that 0 has no reciprocal. The chart shows several rational expressions and their reciprocals. In each case, check that the product of the rational expression and its reciprocal is 1.

Rational Expression	Reciprocal
$\dfrac{4}{x}$	$\dfrac{x}{4}$
$\dfrac{v^2 - 2v}{7}$	$\dfrac{7}{v^2 - 2v}$
$\dfrac{0}{10}$	undefined

The examples in the chart suggest the following procedure.

Procedure: Finding Reciprocals
To find the reciprocal of a nonzero rational expression, invert the rational expression.

Divide Rational Expressions: Dividing rational expressions is like dividing rational numbers.

Procedure: Dividing Rational Expressions
To divide two rational expressions, multiply the first by the reciprocal of the second.

EXAMPLE 4
Divide.

(a) $\dfrac{7z^2}{5} \div \dfrac{3z^5}{10} = \dfrac{7z^2}{5} \cdot \dfrac{10}{3z^5}$ Multiply by the reciprocal of the divisor.

$\qquad\qquad = \dfrac{14}{3z^3}$

(b) $\dfrac{9g-18}{4g} \div \dfrac{4g-8}{7g^2} = \dfrac{9g-18}{4g} \cdot \dfrac{7g^2}{4g-8}$ Multiply by the reciprocal.

$\qquad\qquad = \dfrac{9(g-2)}{4g} \cdot \dfrac{7g^2}{4(g-2)}$ Factor.

$\qquad\qquad = \dfrac{63g}{16}$ Lowest terms

(c) $\dfrac{10x^2-3x-4}{15x^2-47x+28} \div \dfrac{8x^2+22x+9}{9x^2-49} = \dfrac{(5x-4)(2x+1)}{(5x-4)(3x-7)} \div \dfrac{(2x+1)(4x+9)}{(3x-7)(3x+7)}$

$\qquad\qquad = \dfrac{(5x-4)(2x+1)}{(5x-4)(3x-7)} \cdot \dfrac{(3x-7)(3x+7)}{(2x+1)(4x+9)}$

$\qquad\qquad = \dfrac{3x+7}{4x+9}$

EXAMPLE 5

Divide $\dfrac{a^2bc^5}{a^3b^7}$ by $\dfrac{a^7b^2c^8}{a^2bc^3}$.

Use the definitions of division and multiplication and the properties of exponents.

$$\frac{a^2bc^5}{a^3b^7} \div \frac{a^7b^2c^8}{a^2bc^3} = \frac{a^2bc^5}{a^3b^7} \cdot \frac{a^2bc^3}{a^7b^2c^8}$$

$$= \frac{a^4b^2c^8}{a^{10}b^9c^8}$$

$$= \frac{1}{a^6b^7}$$

Self-Check 3

Find the reciprocal of the following.

1. $\dfrac{3}{x}$

2. $\dfrac{x^2-2}{x+1}$

Divide.

3. $\dfrac{3x^2y}{2z} \div \dfrac{6xz^4}{7y^6}$

4. $\dfrac{x^2-16}{x^2-36} \div \dfrac{x^2-2x-8}{x^2+10x+24}$

Self-Check Answers

1.1 $\dfrac{x}{4y^2z}$ 1.2 $\dfrac{x+2}{x-2}$ 1.3 $\dfrac{c-5}{2}$ 1.4 $-(x+4)$ or $-x-4$

2.1 $\dfrac{y^3}{4x}$ 2.2 $\dfrac{3}{4a}$ 2.3 $\dfrac{z-2}{2z^3}$ 2.4 $-20x$

3.1 $\dfrac{x}{3}$ 3.2 $\dfrac{x+1}{x^2+1}$ 3.3 $\dfrac{7xy^7}{4z^5}$ 3.4 $\dfrac{(x+4)^2}{(x-6)(x+2)}$

4.2 EXERCISES

Rational expressions often can be written in lowest terms in seemingly different ways. For example,

$$\frac{x-4}{-7} \quad and \quad \frac{-x+4}{7} \quad and \quad \frac{4-x}{7}$$

look different, but we get the second by multiplying the first by -1 in both the numerator and denominator. The third was obtained by rearranging the terms of the second. As practice in recognizing equivalent rational expressions, match the expressions in exercises 1-5 with their equivalents in choices A-E.

1. $\dfrac{x-2}{x+5}$ A. $\dfrac{-x-2}{5-x}$

2. $\dfrac{x+2}{x-5}$ B. $\dfrac{2-x}{-x-5}$

3. $\dfrac{x-2}{x-5}$ C. $\dfrac{-x+2}{-x+5}$

4. $\dfrac{2-x}{x+5}$ D. $\dfrac{-x-2}{x-5}$

5. $\dfrac{x+2}{5-x}$ E. $\dfrac{x-2}{-x-5}$

6. Only one of the following rational expressions can be simplified. Which one is it?

(a) $\dfrac{x^2+4}{x^2}$ (b) $\dfrac{x^2+4}{4}$ (c) $\dfrac{x^2+y^2}{x^2}$ (d) $\dfrac{x^2-4x}{x}$

Write each rational expression in lowest terms in exercises 7-30.

7. $\dfrac{8x^3y^6}{6x^9y}$

8. $\dfrac{54a^2b^6}{63a^5b^2}$

9. $\dfrac{(x+6)(x-2)}{(x-2)(x+9)}$

10. $\dfrac{(3x-7)(3x+7)}{(x-9)(3x-7)}$

11. $\dfrac{2x^2(x+2)}{4x(x-2)}$

12. $\dfrac{12y^4(y-3)}{20y^3(y+3)}$

13. $\dfrac{2x-7}{2}$

14. $\dfrac{3q+5}{3}$

15. $\dfrac{3x+21}{4x+28}$

16. $\dfrac{6r-4}{15r-10}$

17. $\dfrac{t^2+2t}{6t+12}$

18. $\dfrac{2z^2-8z}{5z-20}$

19. $\dfrac{6y+18}{y^2-9}$

20. $\dfrac{6b-24}{b^2-16}$

21. $\dfrac{c^2-6c+5}{c^2-8c+7}$

22. $\dfrac{a^2-7a+8}{a^2-10a+9}$

23. $\dfrac{4k^2+20k+9}{4k^2-4k-3}$

24. $\dfrac{18p^2-27p+7}{6p^2+7p-3}$

25. $\dfrac{g-2}{g^3-8}$

26. $\dfrac{8x^3+1}{2x+1}$

27. $\dfrac{18x^2+51xy+36y^2}{54x^2+54xy-24y^2}$

28. $\dfrac{30a^2+52ab+16b^2}{18a^2-12ab-48b^2}$

29. $\dfrac{xz-xs+yz-ys}{xz-xs-yz+ys}$

30. $\dfrac{3ab+3ac+b+c}{3ab+b-3ac-c}$

31. Only one of the following rational expression is *not* equivalent to

$$\frac{x-4}{5-x}.$$

Which one is it?

(a) $\dfrac{4-x}{x-5}$ (b) $\dfrac{x+4}{5+x}$ (c) $-\dfrac{4-x}{5-x}$ (d) $-\dfrac{x-4}{x-5}$

32. Which of the following rational expressions equals -1?

(a) $\dfrac{3x+5}{3x-5}$ (b) $\dfrac{3x-5}{5-3x}$ (c) $\dfrac{3x+5}{5+3x}$ (d) $\dfrac{3x+5}{-3x-5}$

Write each rational expression in lowest terms in exercises 29-38.

29. $\dfrac{2-x}{x-2}$

30. $\dfrac{y-4}{4-y}$

31. $\dfrac{a^2-b^2}{b-a}$

32. $\dfrac{s^2-t^2}{t-s}$

33. $\dfrac{(a-2)(3a-1)}{(2-a)(3a+1)}$

34. $\dfrac{(x-5)(2x+3)}{(5-x)(6x-7)}$

35. $\dfrac{6p-12}{8-4p}$

36. $\dfrac{4b-28}{14-2b}$

37. $\dfrac{a^2+b^2}{a^2-b^2}$

38. $\dfrac{x^2-y^2}{x^2+y^2}$

39. Explain in a few words how to multiply rational expressions. Give an example.

40. Explain in a few words how to divide rational expressions. Give an example.

Multiply of divide as indicated in exercises 41-64.

41. $\dfrac{3x^2y}{14} \cdot \dfrac{7y}{6x}$

42. $\dfrac{15a^7}{3b^2} \cdot \dfrac{21b^5}{10a^5}$

43. $\dfrac{30a^5b}{11b^7} \div \dfrac{6b^5}{55a^9}$

44. $\dfrac{49x^8y}{6x^5} \div \dfrac{7y^2}{12x^3y^2}$

45. $\dfrac{\left(2x^2y^3\right)^2}{3x^3} \div \dfrac{8xy^2}{9y^7}$

46. $\dfrac{\left(3a^3b^4\right)^3}{\left(4ab^2\right)^2} \div \dfrac{a^5b}{8ab^6}$

47. $\dfrac{6x}{5x+10} \cdot \dfrac{3x+6}{8}$

48. $\dfrac{2x-12}{10x} \cdot \dfrac{5x^2}{3x-18}$

49. $\dfrac{x^2-16}{5x} \cdot \dfrac{10}{4-x}$

50. $\dfrac{y^2-25}{7y} \cdot \dfrac{14y^2}{5-y}$

51. $\dfrac{d^2-64}{d+5} \div \dfrac{8-d}{d}$

52. $\dfrac{n^2-100}{9n^2} \div \dfrac{10-n}{n}$

53. $\dfrac{17x-51y}{3x+6y} \cdot \dfrac{5x+10y}{6y-2x}$

54. $\dfrac{8s-5t}{21s+27t} \cdot \dfrac{14s+18t}{15t-24s}$

55. $\dfrac{x^2-36}{x-9} \cdot \dfrac{x^2+2x-35}{x^2-11x+30}$

56. $\dfrac{y^2-49}{y+14} \cdot \dfrac{y^2+2y-48}{y^2+15y+56}$

57. $\dfrac{6x^2-xy-12y^2}{3x^2-10xy+3y^2} \div \dfrac{6x^2-11xy+3y^2}{6x^2+xy-y^2}$

58. $\dfrac{2a^2-11ab+15b^2}{2a^2-5ab-3b^2} \div \dfrac{2a^2+ab-15b^2}{a^2-9b^2}$

59. $\dfrac{2s^2+11st+14t^2}{2s^2-9st-56t^2} \div \dfrac{2s^2-5st-18t^2}{s^2-4t^2}$

60. $\dfrac{14c^2-3cd-2d^2}{4c^2-d^2} \div \dfrac{10c^2-cd-2d^2}{10c^2+9cd+2d^2}$

61. $\left(\dfrac{14x^2-9x-8}{24x^2+13x-2} \div \dfrac{28x^2-67x+40}{2x-5}\right) \cdot \dfrac{12x^2-7x-10}{4x^2-8x-5}$

62. $\left(\dfrac{24k+3}{30k^2-86k+16} \div \dfrac{54k^2-288k+384}{7k+1}\right) \cdot \dfrac{30k^2-86k+16}{56k^2+15k+1}$

63. $\dfrac{x^2(2x+1) - 3x(2x+1) - 18(2x+1)}{x^2 - 36} \div \dfrac{4x^2 - 1}{14x^2 - 5x - 1}$

64. $\dfrac{a^2(3a+2b) - 2ab(3a+2b) - 8b^2(3a+2b)}{a^2 - 16b^2} \div \dfrac{9a^2 - 4b^2}{15a^2 - 4ab - 4b^2}$

| 4.3 | **Addition and Subtraction of Rational Expressions** |

Add and Subtract Rational Expressions with the Same Denominator: The following steps, used to add or subtract rational numbers, are also used to add or subtract rational expressions.

Procedure: Adding or Subtracting Rational Expressions

Step 1 **If the denominators are the same,** add or subtract the numerators. Place the result over the common denominator.

If the denominators are different, first find the least common denominator. Write all rational expressions with this least common denominator, and then add or subtract the numerators. Place the result over the common denominator

Step 2 **Simplify.** Write all answers in lowest terms.

EXAMPLE 1

Add or subtract as indicated.

(a) $\dfrac{3x}{7} + \dfrac{2y}{7} = \dfrac{3x + 2y}{7}$

The denominators of these rational expressions are the same, so just add the numerators, and place the sum over the common denominator.

(b) $\dfrac{19x}{3z^2} - \dfrac{x}{3z^2} = \dfrac{19x - x}{3z^2} = \dfrac{18x}{3z^2} = \dfrac{6x}{z^2}$ Lowest terms

Subtract the numerators since the denominators are the same, and keep the common denominator.

(c) $\dfrac{x}{x^2 - y^2} + \dfrac{y}{x^2 - y^2} = \dfrac{x + y}{x^2 - y^2}$ Add numerators, keep the common denominator

$= \dfrac{x + y}{(x - y)(x + y)}$ Factor.

$= \dfrac{1}{x - y}$ Lowest terms

(d) $\dfrac{d}{d^2 - 3d - 18} + \dfrac{3}{d^2 - 3d - 18} = \dfrac{d + 3}{d^2 - 3d - 18}$

$= \dfrac{d + 3}{(d - 6)(d + 3)}$

$= \dfrac{1}{d - 6}$

Self-Check 1

Add or subtract as indicated

1. $\dfrac{8a}{11} + \dfrac{9b}{11}$

2. $\dfrac{15g}{2k^3} - \dfrac{g}{2k^3}$

3. $\dfrac{c}{c^2 - 4} + \dfrac{2}{c^2 - 4}$

4. $\dfrac{x}{x^2 + 2x - 8} - \dfrac{2}{x^2 + 2x - 8}$

Find a Least Common Denominator: We add or subtract rational expressions with different denominators by first writing them with a common denominator, usually the **least common denominator (LCD)**.

Procedure: Finding the Least Common Denominator

Step 1 **Factor.** Factor each denominator.

Step 2 **Find the least common denominator.** The LCD is the product of all different factors from each denominator, with each factor raised to the *greatest* power that occurs in any denominator.

E X A M P L E 2

Find the least common denominator for each pair of denominators.

(a) $3x^2y, \ 7xy^3$

Each denominator is already factored.

$$3x^2y = 3 \cdot x^2 \cdot y$$
$$7xy^3 = 7 \cdot x \cdot y^3$$

Greatest exponent on x is 2.

$$\text{LCD} = 3 \cdot 7 \cdot x^2 \cdot y^3 \quad \longleftarrow \text{Greatest exponent on } y \text{ is 3.}$$
$$= 21x^2y^3$$

(b) $z - 2, \ z$

The LCD, an expression divisible by both $z - 2$ and z, is

$$z(z - 2).$$

It is often best to leave a least common denominator in factored form.

(c) $y^2 - 5y - 6, \ y^2 + 2y + 1$

Factor the denominators to get

$$y^2 - 5y - 6 = (y - 6)(y + 1)$$
$$y^2 + 2y + 1 = (y + 1)^2.$$

The LCD, divisible by both polynomials, is

$$(y - 6)(y + 1)^2.$$

Self-Check 2

Find the least common denominator for each pair of denominators.

1. 36, 24

2. $6x^2yz^5$, $8x^3z$

3. $3x - 6$, $x^2 + 4x - 12$

4. $2y^2 + 2y - 24$, $6y^2 - 36y + 54$

Add and Subtract Rational Expressions with Different Denominators: Before adding or subtracting two rational expressions, we write each expression with the least common denominator by multiplying its numerator and denominator by the factors needed to get the LCD. This procedure is valid because we are multiplying each rational expression by a form of 1, the identity element for multiplication.

EXAMPLE 3

Add or subtract as indicated.

(a) $\dfrac{5}{12w} + \dfrac{9}{16w}$

The LCD for $12w$ and $16w$ is $48w$. To write the first and the second rational expressions with a denominator of $48w$, multiply by $\frac{4}{4}$ and $\frac{3}{3}$ respectively.

$$\frac{5}{12w} + \frac{9}{16w} = \frac{5 \cdot 4}{12w \cdot 4} + \frac{9 \cdot 3}{16w \cdot 3} \quad \text{Fundamental principle}$$

$$= \frac{20}{48w} + \frac{27}{48w}$$

$$= \frac{20 + 27}{48w} \quad \text{Add numerators.}$$

$$= \frac{47}{48w}$$

(b) $\dfrac{3}{z} - \dfrac{2}{z - 5}$

The LCD is $z(z - 5)$. Rewrite each rational expression with this denominator.

$$\frac{3}{z} - \frac{2}{z - 5} = \frac{3(z - 5)}{z(z - 5)} - \frac{z \cdot 2}{z(z - 5)} \quad \text{Fundamental principle}$$

$$= \frac{3z - 15}{z(z - 5)} - \frac{2z}{z(z - 5)} \quad \text{Distributive property}$$

$$= \frac{3z - 15 - 2z}{z(z - 5)} \quad \text{Subtract numerators.}$$

$$= \frac{z - 15}{z(z - 5)} \quad \text{Combine terms in numerator.}$$

E X A M P L E   4

Subtract.

(a) $\dfrac{4}{r+1} - \dfrac{r-2}{r-3}$

The LCD is $(r+1)(r-3)$.

$$\dfrac{4}{r+1} - \dfrac{r-2}{r-3} = \dfrac{4(r-3)}{(r+1)(r-3)} - \dfrac{(r-2)(r+1)}{(r-3)(r+1)} \qquad \text{Fundamental principle}$$

$$= \dfrac{4r-12}{(r+1)(r-3)} - \dfrac{r^2-r-2}{(r-3)(r+1)} \qquad \text{Multiply in numerator.}$$

$$= \dfrac{4r-12-\left(r^2-r-2\right)}{(r+1)(r-3)} \qquad \text{Subtract.}$$

$$= \dfrac{4r-12-r^2+r+2}{(r+1)(r-3)} \qquad \text{Distributive property}$$

$$= \dfrac{-r^2+5r-10}{(r+1)(r-3)} \qquad \text{Combine terms.}$$

(b) $\dfrac{2}{x-2} - \dfrac{2}{x+2} = \dfrac{2(x+2)}{(x-2)(x+2)} - \dfrac{2(x-2)}{(x+2)(x-2)} \qquad \text{Get a common denominator.}$

$$= \dfrac{2x+4}{(x-2)(x+2)} - \dfrac{2x-4}{(x-2)(x+2)} \qquad \text{Multiply in numerators.}$$

$$= \dfrac{2x+4-(2x-4)}{(x-2)(x+2)} \qquad \text{Subtract.}$$

$$= \dfrac{2x+4-2x+4}{(x-2)(x+2)} \qquad \text{Distributive property}$$

$$= \dfrac{8}{(x-2)(x+2)} \qquad \text{Combine terms.}$$

Self-Check 3

Add or subtract as indicated.

1. $\dfrac{3}{2c} + \dfrac{7}{5c^2}$

2. $\dfrac{5}{x} - \dfrac{9}{x-6}$

3. $\dfrac{5}{f-7} + \dfrac{f+1}{f-4}$

4. $\dfrac{5}{x-7} - \dfrac{7}{x+5}$

When you are adding rational expressions that have opposites in the denominator, multiply numerator and denominator of that term by -1. By the fundamental principle the value does not change, just its appearance.

EXAMPLE 5

$$\frac{y}{(y-2)^2} + \frac{3y}{4-y^2} = \frac{y}{(y-2)^2} + \frac{3y}{(2-y)(2+y)}$$ Factor denominators.

$$= \frac{y}{(y-2)^2} + \frac{-1 \cdot 3y}{-1(2-y)(2+y)}$$ $y-2$ and $2-y$
are opposites.

$$= \frac{y}{(y-2)^2} + \frac{-3y}{(y-2)(y+2)}$$

$$= \frac{y(y+2)}{(y-2)^2(y+2)} + \frac{-3y(y-2)}{(y-2)^2(y+2)}$$

$$= \frac{y^2+2y}{(y-2)^2(y+2)} + \frac{-3y^2+6y}{(y-2)^2(y+2)}$$ Multiply in numerators.

$$= \frac{y^2+2y-3y^2+6y}{(y-2)^2(y+2)}$$ Add numerators.

$$= \frac{-2y^2+8y}{(y-2)^2(y+2)}$$ Combine terms.

$$= \frac{-2y(y-4)}{(y-2)^2(y+2)}$$ Factor numerator for
possible reduction.

Self-Check 4

Add or subtract as indicated.

1. $\dfrac{3}{b^2-2b-8} + \dfrac{2b}{16-b^2}$ **2.** $\dfrac{g}{(g-3)^2} - \dfrac{2}{9-g^2}$

Self-Check Answers

1.1 $\dfrac{8a+9b}{11}$ **1.2** $\dfrac{7g}{k^3}$ **1.3** $\dfrac{1}{c-2}$ **1.4** $\dfrac{1}{x+4}$

2.1 72 **2.2** $24x^3yz^5$ **2.3** $3(x-2)(x+6)$ **2.4** $6(y+4)(y-3)^2$

3.1 $\dfrac{15c+14}{10c^2}$ **3.2** $\dfrac{-4x-30}{x(x-6)}$ **3.3** $\dfrac{f^2-f-27}{(f-7)(f-4)}$ **3.4** $\dfrac{-2x+74}{(x-7)(x+5)}$

4.1 $\dfrac{-2b^2-b+12}{(b-4)(b+2)(b+4)}$ **4.2** $\dfrac{(g+6)(g-1)}{(g-3)^2(g+3)}$

4.3 EXERCISES

1. Let $x = 5$ and $y = 4$. Evaluate $\dfrac{1}{x} + \dfrac{1}{y}$.

2. Let $x = 5$ and $y = 4$. Evaluate $\dfrac{1}{x+y}$.

3. Are the answers for Exercises 1 and 2 the same? What can you conclude?

4. Let $x = 7$ and $y = 8$. Evaluate $\dfrac{1}{x} - \dfrac{1}{y}$.

5. Let $x = 7$ and $y = 8$. Evaluate $\dfrac{1}{x - y}$.

6. Are the answers for Exercises 4 and 5 the same? What can you conclude?

Add or subtract as indicated in exercises 7-18. Write the answer in lowest terms.

7. $\dfrac{2}{c} + \dfrac{3}{c}$

8. $\dfrac{4}{z} + \dfrac{8}{z}$

9. $\dfrac{13}{2x^2} - \dfrac{5}{2x^2}$

10. $\dfrac{17}{3ab^2} + \dfrac{1}{3ab^2}$

11. $\dfrac{3x - 7}{4x - 5} + \dfrac{x + 2}{4x - 5}$

12. $\dfrac{7b - 6}{3b + 1} - \dfrac{b - 8}{3b + 1}$

13. $\dfrac{d^2}{d + 4} - \dfrac{16}{d + 4}$

14. $\dfrac{z^2}{z - 7} - \dfrac{49}{z - 7}$

15. $\dfrac{-3x + 7}{x^2 + 9x + 8} + \dfrac{8x + 33}{x^2 + 9x + 8}$

16. $\dfrac{z - 6}{z^2 - 5z + 6} + \dfrac{2z - 3}{z^2 - 5z + 6}$

17. $\dfrac{a^3}{a^2 + 2a + 4} - \dfrac{8}{a^2 + 2a + 4}$

18. $\dfrac{p^3}{p^2 - 3pq + 9q^2} + \dfrac{27q^3}{p^2 - 3pq + 9q^2}$

Find the least common denominator for each group of denominators in exercises 19-38.

19. $16x^2y^3, 24xy^9z^6$

20. $42a^3b^2, 36a^3b^5c^7$

21. $c - 7, c$

22. $d + 4, d$

23. $4x + 16, x + 4$

24. $3y - 15, y - 5$

25. $6p - 12, 8p - 16$

26. $10r + 50, 15r + 75$

27. $a - b, a + b$

28. $x + y, x - y$

29. $\dfrac{y - 3}{y^2 - 6y - 7}, \dfrac{y + 9}{y^2 + y}$

30. $\dfrac{x + 5}{x^2 + 2x - 8}, \dfrac{-6}{x^2 - 2x}$

31. $\dfrac{p - 3}{p^2 - 3p - 18}, \dfrac{-9p}{p^2 + 3p}$

32. $\dfrac{d + 2}{d^2 - 7d + 12}, \dfrac{d}{d^2 - 4d}$

33. $\dfrac{s}{3s^2 + s - 10}, \dfrac{s}{s^2 - 3s - 10}$

34. $\dfrac{x}{2x^2 - 11x + 12}, \dfrac{x}{x^2 - x - 12}$

35. $\dfrac{3}{2x - 4}, \dfrac{4x}{x^2 - 4}, \dfrac{9}{x}$

36. $\dfrac{7r}{3r + 9}, \dfrac{r}{r^2 - 9}, \dfrac{2}{r}$

37. $\dfrac{2}{g^2}, \dfrac{7}{4g^5}, \dfrac{11}{6g - 12}$

38. $\dfrac{1}{p^3}, \dfrac{2}{5p}, \dfrac{3p}{10p + 20}$

39. One student added two rational expressions and obtained the answer $-\dfrac{2}{x-3}$.

Another student obtained the answer $\dfrac{2}{3-x}$ for the same problem. Is it possible that both answers are correct? Explain.

40. What is *wrong* with the following work?

$$\frac{3x}{2x+1}-\frac{2x-9}{2x+1}=\frac{3x-2x-9}{2x+1}=\frac{x-9}{2x+1}$$

Add or subtract as indicated in exercises 41-68. Write the answer in lowest terms.

41. $\dfrac{2}{t}+\dfrac{3}{4t}$

42. $\dfrac{3}{x}+\dfrac{9}{5x}$

43. $\dfrac{5}{6x^2y}-\dfrac{7}{8xy^2}$

44. $\dfrac{11}{12a^2b^5}-\dfrac{9}{16a^3b}$

45. $\dfrac{4}{t-4}-\dfrac{4}{t}$

46. $\dfrac{5}{s-5}-\dfrac{5}{s}$

47. $\dfrac{2a}{a-2}+\dfrac{3a}{a-3}$

48. $\dfrac{3x}{x-4}+\dfrac{4x}{x-5}$

49. $\dfrac{2d-1}{d-6}+\dfrac{d+3}{6-d}$

50. $\dfrac{3g-5}{g-2}+\dfrac{g+4}{2-g}$

51. $\dfrac{3a+2b}{a-b}-\dfrac{a+4b}{b-a}$

52. $\dfrac{4x-y}{x-y}-\dfrac{2x-3y}{y-x}$

53. $\dfrac{x}{x-3}+\dfrac{2}{x+3}-\dfrac{18}{x^2-9}$

54. $\dfrac{t}{t-4}+\dfrac{2}{t+4}-\dfrac{32}{t^2-16}$

55. $\dfrac{5}{f-3}+\dfrac{2}{f}-\dfrac{15}{f^2-3f}$

56. $\dfrac{1}{y-5}+\dfrac{9}{y}-\dfrac{5}{y^2-5y}$

57. $\dfrac{2x}{x+2}+\dfrac{1}{x-2}-\dfrac{4}{x^2-4}$

58. $\dfrac{3z}{z+4}+\dfrac{z+1}{z}-\dfrac{4}{z^2+4z}$

59. $\dfrac{2}{v+2}-\dfrac{2}{v^2-2v+4}-\dfrac{24}{v^3+8}$

60. $\dfrac{1}{y-3}-\dfrac{2}{y^2+3y+9}-\dfrac{27}{y^3-27}$

61. $\dfrac{x+1}{x+9}-\dfrac{2}{x}-\dfrac{72}{x^2+9x}$

62. $\dfrac{t+2}{t-4}-\dfrac{3}{t}-\dfrac{12}{t^2-4t}$

63. $\dfrac{6}{(c-3)^2}-\dfrac{6}{c-3}+4$

64. $\dfrac{2}{(2w+1)^2}-\dfrac{3}{2w+1}-1$

65. $\dfrac{x+1}{x^2-3xy+2y^2}+\dfrac{y+1}{x^2+4xy-5y^2}$ 66. $\dfrac{a-2}{a^2-6ab+8b^2}+\dfrac{b+3}{a^2-2ab-6b^2}$

67. $\dfrac{a+b}{2a^2+ab-3b^2}-\dfrac{2a+b}{a^2-b^2}$ 68. $\dfrac{s+t}{3s^2-st-4t^2}-\dfrac{3s+2t}{s^2-t^2}$

4.4 Complex Fractions

A **complex fraction** is an expression having a fraction in the numerator, denominator, or both. Examples of complex fractions include

$$\dfrac{\dfrac{1}{12}+\dfrac{5}{6}}{\dfrac{2}{3}},\quad \dfrac{\dfrac{3}{x}}{1-\dfrac{1}{y}},\quad \text{and}\quad \dfrac{\dfrac{g^2-4}{g+7}}{\dfrac{g-2}{g^2-49}}.$$

Simplify Complex Fractions by Simplifying Numerator and Denominator: There are two different methods for simplifying complex fractions. In the first method, we simplify the numerator and denominator of the complex fraction.

Procedure: Simplifying Complex Fractions: *Method 1*

Step 1 Simplify the numerator and denominator separately, as much as possible.

Step 2 Divide by multiplying the numerator by the reciprocal of the denominator.

Step 3 Simplify the resulting fraction, if possible.

In Step 2, we are treating the complex fraction as a quotient of two rational expressions and dividing. Before performing this step, be sure that both the numerator and denominator are single fractions.

EXAMPLE 1

(a) $\dfrac{\dfrac{x-3}{6x}}{\dfrac{x+5}{8x^2}}$

Both the numerator and the denominator are already simplified, so multiply the numerator by the reciprocal of the denominator.

$$\dfrac{\dfrac{x-3}{6x}}{\dfrac{x+5}{8x^2}}=\dfrac{x-3}{6x}\div\dfrac{x+5}{8x^2} \qquad \text{Write as a division problem.}$$

$$=\dfrac{x-3}{6x}\cdot\dfrac{8x^2}{x+5} \qquad \text{Reciprocal of } \tfrac{x+5}{8x^2}$$

$$=\dfrac{4x(x-3)}{3(x+5)}$$

(b) $\dfrac{5 - \dfrac{1}{p}}{2 + \dfrac{3}{p}} = \dfrac{\dfrac{5p}{p} - \dfrac{1}{p}}{\dfrac{2p}{p} + \dfrac{3}{p}} = \dfrac{\dfrac{5p - 1}{p}}{\dfrac{2p + 3}{p}}$ Simplify numerator and denominator.

$\qquad = \dfrac{5p - 1}{p} \cdot \dfrac{p}{2p + 3}$ Reciprocal of $\frac{2p+3}{p}$

$\qquad = \dfrac{5p - 1}{2p + 3}$

Self-Check 1

Simplify. Assume all variables represent nonzero real numbers.

1. $\dfrac{\dfrac{a + 3}{8a}}{\dfrac{a - 4}{4a^2}}$

2. $\dfrac{\dfrac{1}{x} + \dfrac{3}{x^2}}{\dfrac{7}{x} - 1}$

Simplify Complex Fractions by Multiplying by a Common Denominator: The second method for simplifying complex fractions uses the identity property for multiplication.

Procedure: Simplifying Complex Fractions: *Method 2*

Step 1 Multiply the numerator and denominator of the complex fraction by the least common denominator of the fractions in the numerator and the fractions in the denominator of the complex fraction.

Step 2 Simplify the resulting fraction, if possible.

EXAMPLE 2

(a) $\dfrac{5 - \dfrac{1}{p}}{2 + \dfrac{3}{p}}$

Multiply the numerator and denominator by the LCD of all the fractions in the numerator and the denominator of the complex fraction. (This is the same as multiplying by 1.) Here the LCD is p.

$\dfrac{5 - \dfrac{1}{p}}{2 + \dfrac{3}{p}} = \dfrac{5 - \dfrac{1}{p}}{2 + \dfrac{3}{p}} \cdot 1 = \dfrac{\left(5 - \dfrac{1}{p}\right) \cdot p}{\left(2 + \dfrac{3}{p}\right) \cdot p}$ Multiply numerator and denominator by p, since $\frac{p}{p} = 1$.

$\qquad = \dfrac{5 \cdot p - \dfrac{1}{p} \cdot p}{2 \cdot p + \dfrac{3}{p} \cdot p}$ Use the distributive property.

$\qquad = \dfrac{5p - 1}{2p + 3}$

(b) $\dfrac{2z - \dfrac{3}{z+1}}{3z + \dfrac{5}{z}}$

The LCD is $z(z + 1)$.

$$\frac{2z - \dfrac{3}{z+1}}{3z + \dfrac{5}{z}} = \frac{2z\big[z(z+1)\big] - \dfrac{3}{z+1} \cdot z(z+1)}{3z\big[z(z+1)\big] + \dfrac{5}{z} \cdot z(z+1)}$$

$$= \frac{2z\big[z(z+1)\big] - 3z}{3z\big[z(z+1)\big] + 5(z+1)}$$

$$= \frac{2z^3 + 2z^2 - 3z}{3z^3 + 3z^2 + 5z + 5}$$

Self-Check 2

Simplify using Method 2. Assume all variables represent nonzero real numbers.

1. $\dfrac{m + \dfrac{5}{n}}{3n - \dfrac{m}{n}}$

2. $\dfrac{5b + \dfrac{2}{b+3}}{5b^2 - \dfrac{7}{b}}$

Simplify Rational Expressions with Negative Exponents: Rational expressions and complex fractions often involve negative exponents. To simplify such expressions, we begin by rewriting the expressions with only positive exponents.

EXAMPLE 3
Simplify each of the following.

(a) $\dfrac{x^{-1} + 5y^{-3}}{7x^{-2} - y^{-4}}$

First write the expression with only positive exponents using the definition of a negative exponent.

$$\frac{x^{-1} + 5y^{-3}}{7x^{-2} - y^{-4}} = \frac{\dfrac{1}{x} + \dfrac{5}{y^3}}{\dfrac{7}{x^2} - \dfrac{1}{y^4}}$$

Note that the 5 in $5y^{-3}$ is not raised to the -3 power, so $5y^{-3} = \dfrac{5}{y^3}$. Simplify the complex fraction by multiplying numerator and denominator by the LCD, $x^2 y^4$.

$$\frac{\dfrac{1}{x} + \dfrac{5}{y^3}}{\dfrac{7}{x^2} - \dfrac{1}{y^4}} = \frac{x^2 y^4 \cdot \dfrac{1}{x} + x^2 y^4 \cdot \dfrac{5}{y^3}}{x^2 y^4 \cdot \dfrac{7}{x^2} - x^2 y^4 \cdot \dfrac{1}{y^4}} \qquad \text{Use Method 2.}$$

$$= \frac{xy^4 + 5x^2 y}{7y^4 - x^2} \text{ or } \frac{xy\big(y^3 + 5x\big)}{7y^4 - x^2}$$

(b) $\dfrac{k^{-2} - 1}{k^{-2} + k^{-1}} = \dfrac{\dfrac{1}{k^2} - 1}{\dfrac{1}{k^2} + \dfrac{1}{k}}$ Write with positive exponents.

$$= \dfrac{k^2 \cdot \dfrac{1}{k^2} - k^2 \cdot 1}{k^2 \cdot \dfrac{1}{k^2} + k^2 \cdot \dfrac{1}{k}}$$ Use Method 2.

$$= \dfrac{1 - k^2}{1 + k}$$

$$= \dfrac{(1 - k)(1 + k)}{1 + k}$$

$$= 1 - k$$

Self-Check 3

Simplify each of the following.

1. $\dfrac{a^{-2} - 2b^{-4}}{6a^{-1} + b^{-6}}$ 2. $\dfrac{s^{-4} - 5}{s^{-3} - 2s^{-1}}$

Self-Check Answers

1.1 $\dfrac{a(a + 3)}{2(a - 4)}$ 1.2 $\dfrac{x + 3}{7x - x^2}$ or $\dfrac{x + 3}{x(7 - x)}$

2.1 $\dfrac{mn + 5}{3n^2 - m}$ 2.2 $\dfrac{5b^3 + 15b^2 + 2b}{5b^4 + 15b^2 - 7b - 21}$

3.1 $\dfrac{b^6 - 2a^2b^2}{6ab^6 + a^2}$ or $\dfrac{b^2(b^4 - 2a^2)}{a(6b^6 + a)}$ 3.2 $\dfrac{1 - 5s^4}{s - 2s^3}$ or $\dfrac{1 - 5s^4}{s(1 - 2s^2)}$

4.4 EXERCISES

1. Explain in your own words the two methods of simplifying complex fractions.

2. Method 2 of simplifying complex fractions says that we can multiply both the numerator and the denominator of the complex fraction by the same nonzero expression. What property of real numbers justifies this method?

Use either method to simplify each complex fraction in exercises 3-20.

3. $\dfrac{\dfrac{24}{x - 5}}{\dfrac{16}{x}}$ 4. $\dfrac{\dfrac{36}{t + 2}}{\dfrac{20}{t}}$

5. $\dfrac{\dfrac{3z - 8}{4z}}{\dfrac{2z + 1}{8z}}$ 6. $\dfrac{\dfrac{9p - 7}{14p^2}}{\dfrac{3p + 4}{7p^2}}$

7. $\dfrac{\dfrac{3x^2y}{20}}{\dfrac{6xy^3}{25}}$

8. $\dfrac{\dfrac{55ab^6}{48}}{\dfrac{11a^2b^5}{24}}$

9. $\dfrac{\dfrac{3}{d}-1}{2+\dfrac{5}{d}}$

10. $\dfrac{\dfrac{2}{p}+\dfrac{3}{p}}{1-\dfrac{7}{p}}$

11. $\dfrac{\dfrac{2}{x}+\dfrac{2}{y}}{\dfrac{2}{x}-\dfrac{2}{y}}$

12. $\dfrac{\dfrac{5}{m}-\dfrac{5}{n}}{\dfrac{5}{m}+\dfrac{5}{n}}$

13. $\dfrac{\dfrac{2x-4y}{12}}{\dfrac{7x-14y}{4x}}$

14. $\dfrac{\dfrac{2m-n}{12m}}{\dfrac{6m-3n}{4m^2}}$

15. $\dfrac{\dfrac{a^2-9b^2}{ab}}{\dfrac{1}{b}-\dfrac{3}{a}}$

16. $\dfrac{\dfrac{4}{m}+\dfrac{3}{n}}{\dfrac{9m^2-16n^2}{mn}}$

17. $\dfrac{d-\dfrac{d-5}{3}}{\dfrac{4}{15}+\dfrac{2}{3d}}$

18. $\dfrac{3x-\dfrac{x-16}{7}}{\dfrac{5}{14}+\dfrac{2}{7x}}$

19. $\dfrac{\dfrac{z+4}{z}-\dfrac{8}{z-3}}{\dfrac{3}{z}+\dfrac{z}{z-3}}$

20. $\dfrac{\dfrac{y-3}{y}+\dfrac{2}{y+2}}{\dfrac{3}{y+2}-\dfrac{2}{y}}$

Simplify each expression, using only positive exponents in your answer in exercises 21-26.

21. $\dfrac{1}{x^{-4}+y^{-4}}$

22. $\dfrac{1}{p^{-6}+q^{-6}}$

23. $\dfrac{x^{-2}-y^{-2}}{x^{-1}-y^{-1}}$

24. $\dfrac{m^{-1}+n^{-1}}{m^{-2}+n^{-2}}$

25. $\left(x^{-1}-y^{-1}\right)^{-1}$

26. $\left[(2a)^{-1}+(3b)^{-1}\right]^{-1}$

CH 4   Summary

KEY TERMS

4.1 factored
  prime
  lowest terms
  basic principle of fractions
  least common denominator (LCD) of fractions

greatest common factor
reciprocal
quotient

4.2 rational expression   fundamental principle of rational numbers

4.3 least common denominator (LCD) of rational expressions

4.4 complex fraction

CH 4   Quick Review

4.1 FRACTIONS

Operations with Fractions

Addition/Subtraction

1. To add/subtract fractions with the same denominator, add/subtract the numerators and keep the same denominator.
2. To add/subtract fractions with different denominators, find the LCD and write each fraction with this LCD. Then follow the procedure above.

Multiplication: Multiply numerators and multiply denominators.

Division: Multiply the first fraction by the reciprocal of the second fraction.

4.2 RATIONAL EXPRESSIONS: MULTIPLICATION AND DIVISION

Writing Rational Expressions in Lowest Terms

Step 1   Factor the numerator and denominator completely

Step 2   Apply the fundamental principle.

Multiplying Rational Expressions

Factor numerators and denominators. Apply the fundamental principle; replace all pairs of common factors in numerators and denominators by 1. Multiply remaining factors in the numerator and in the denominator.

Dividing Rational Expressions

Multiply the first fraction by the reciprocal of the second.

4.3 ADDITION AND SUBTRACTION OF RATIONAL EXPRESSIONS

Adding or Subtracting Rational Expressions

If the denominators are the same, add or subtract the numerators. Place the result over the common denominator. If the denominators are different, write all rational expressions with the LCD. Then add or subtract the like fractions. Be sure the final answer is in lowest terms.

4.4 COMPLEX FRACTIONS

Simplifying Complex Fractions

Method 1 Simplify the numerator and denominator separately, as much as possible. Then multiply the numerator by the reciprocal of the denominator. Write the answer in lowest terms.

Method 2 Multiply the numerator and denominator of the complex fraction by the least common denominator of all fractions in the complex fraction. Then simplify the results.

CH 4   Review Exercises

Write each fraction in lowest terms in the following exercise.

1. a) $\dfrac{26}{39}$ b) $\dfrac{192}{256}$ c) $\dfrac{126}{306}$

Find each product or quotient, and write in lowest terms in the following exercise.

2. a) $\dfrac{3}{4} \cdot \dfrac{5}{7}$ b) $\dfrac{1}{3} \div \dfrac{5}{12}$ c) $\dfrac{2}{7} \div \dfrac{14}{5}$

Find each sum or difference, and write in lowest terms in the following exercise.

3. a) $\dfrac{1}{12} + \dfrac{2}{12}$ b) $\dfrac{15}{16} - \dfrac{1}{24}$ c) $\dfrac{5}{6} - \dfrac{1}{4} + \dfrac{5}{12}$

4. List the steps you would use to write $\dfrac{x^2 - 6x}{2x - 12}$ in lowest terms.

Write in lowest terms in exercises 5-8.

5. $\dfrac{56x^3 y^2}{49x^2 y^6}$ **6.** $\dfrac{36x^2 - 6x}{12xy - 2y}$

7. $\dfrac{10a^2 + 63ab + 81b^2}{10a^2 + 3ab - 27b^2}$ **8.** $\dfrac{10 - 5r}{r^2 - 4}$

9. What is meant by the reciprocal of a rational expression?

Multiply or divide. Write the answer in lowest terms in exercises 10-13.

10. $\dfrac{26st^8}{3st} \div \dfrac{13s^2 t}{6s^5}$ **11.** $\dfrac{g^2 - 25}{g} \cdot \dfrac{5}{5 - g}$

12. $\dfrac{x^2 - 3x - 18}{x^2 - 2x - 3} \cdot \dfrac{x^2 + 7x + 6}{x^2 - 36}$ **13.** $\dfrac{p^2 - q^2}{p^3 - q^3} \div \dfrac{p + q}{p^2 + pq + q^2}$

14. What is *wrong* with the following work?

$$\frac{d^2 + 3d}{d + 3} = \frac{d^2}{d} + \frac{3d}{3} = d + 3$$

What is the *correct* simplified form?

15. How would you explain finding the least common denominator to a classmate?

Find the least common denominator for each group of rational expressions in exercises 16-18.

16. $\dfrac{5g}{16h^3}, \dfrac{7}{24h^2}$ **17.** $\dfrac{1}{16f^2}, \dfrac{2f - 1}{8f + 4}$ **18** $\dfrac{2x + 5}{x^2 - 49}, \dfrac{3x + 2}{x^2 - 8x + 7}$

Add or subtract as indicated in exercises 19-22.

19. $\dfrac{9}{4c^2} - \dfrac{1}{c}$

20. $\dfrac{4r-2}{r-5} - \dfrac{2r+8}{r-5}$

21. $\dfrac{1}{4a+6} + \dfrac{4}{10a+15}$

22. $\dfrac{6}{8x^2 - 18xy + 9y^2} - \dfrac{3}{8x^2 + 30xy - 27y^2}$

Simplify each complex fraction in exercises 23-25.

23. $\dfrac{\dfrac{5}{b} + 2}{\dfrac{2}{b} - 5}$

24. $\dfrac{\dfrac{m^2 - 25}{m^2 n^4}}{\dfrac{m-5}{mn^5}}$

25. $\dfrac{\dfrac{5}{x} - \dfrac{7}{y}}{\dfrac{25y^2 - 49x^2}{xy}}$

CH 4 | Test

Write each fraction in lowest terms in the following exercise.

1.   a) $\dfrac{126}{238}$

b) $\dfrac{126}{234}$

Find each product or quotient, and write in lowest terms in the following exercise.

2.   a) $\dfrac{9}{55} \cdot \dfrac{25}{39}$

b) $\dfrac{35}{6} \div \dfrac{7}{18}$

Find each sum or difference, and write in lowest terms in the following exercise.

3.   a) $\dfrac{5}{18} + \dfrac{11}{42}$

b) $\dfrac{19}{20} - \dfrac{5}{12}$

Write in lowest terms in exercises 4-7.

4. $\dfrac{18x^3 y - 30x^2 y^2}{6x^2 - 10xy}$

5. $\dfrac{12x^2 + 2x - 2}{8x^2 - 2}$

6. $\dfrac{12a^2 + 50ab - 18b^2}{24a^2 + 4ab - 4b^2}$

7. $\dfrac{5 - 45g^2}{9g^2 + 18g - 7}$

Multiply or divide in exercises 8-11. Write the answer in lowest terms.

8. $\dfrac{3x-6}{14} \cdot \dfrac{7}{2x-4}$

9. $\dfrac{2f+16}{f^2 - 81} \div \dfrac{3f+24}{f^2 - 10f + 9}$

10. $\dfrac{45z^2 + 4z - 1}{1 - 81z^2} \div \dfrac{25z^2 - 4}{45z^2 + 23z + 2}$

11. $\dfrac{6x^2 + 13x + 5}{21x^2 + 47x + 20} \cdot \dfrac{16 - 49x^2}{14x^2 - x - 4}$

Find the least common denominator for each group of rational expressions in exercises 12-14.

12. $\dfrac{5}{2x^2y^7}, \dfrac{7}{12x^3y^5}$

13. $\dfrac{6r^2+13}{6r^3-6r}, \dfrac{2r+11}{16r^2-8r-8}$

14. $\dfrac{2x-9}{18x^2+39x+15}, \dfrac{3x-4}{36x^2+36x+9}$

Add or subtract in exercises 15-18. Write the answer in lowest terms.

15. $\dfrac{13z-2}{4z+7} - \dfrac{7z-6}{4z+7}$

16. $\dfrac{3n}{n^2+2n-8} + \dfrac{5n}{2n^2-4n}$

17. $\dfrac{3x}{x^2-4x+4} - \dfrac{9x+4}{3x^2-5x-2}$

18. $\dfrac{5}{p-1} + \dfrac{1}{p+8} - \dfrac{45p}{p^2+7p-8}$

Simplify each complex fraction in exercises 19-20.

19. $\dfrac{\dfrac{5}{2d+4}}{\dfrac{20}{3d+6}}$

20. $\dfrac{\dfrac{2}{b}-\dfrac{3}{a}}{\dfrac{4a^2-9b^2}{ab^2}}$

CHAPTER 5: ROOTS AND RADICALS; COMPLEX NUMBERS

5.1 | Evaluating Roots

In Chapter 1 we discussed the idea of the *square* of a number. Recall that squaring a number means multiplying the number by itself.

$$\text{If } a = 5, \text{ then } a^2 = 5 \cdot 5 = 25.$$

In this chapter the opposite problem is considered.

$$\text{If } a^2 = 5 \cdot 5 = 25, \text{ then } a = 5.$$

To find a in the statement above, we must find a number that when multiplied by itself results in the given number. The number a is called a **square root** of the number a^2.

Find All Square Roots of a Number: To find a square root of a number like 25, think of a number that when multiplied by itself gives 25. One square root is 5, since $5 \cdot 5 = 25$. Another square root of 25 is -5, since $(-5)(-5) = 25$. The number 25 has two square roots, 5 and -5; one is positive and one is negative.

The positive square root of a number is written with the symbol $\sqrt{}$. For example, the positive square root of 100 is 10, written

$$\sqrt{100} = 10.$$

The symbol $-\sqrt{}$ is used for the negative square root of a number. For example, the negative square root of 100 is -10, written

$$-\sqrt{100} = -10.$$

Most calculators have a square root key, usually labeled $\boxed{\sqrt{x}}$, that allows us to find the square root of a number. For example, if we enter 100 and use the square root key, the display will show 10.

The symbol $\sqrt{}$ is called a **radical sign** and always represents the nonnegative square root. The number inside the radical sign is called the **radicand** and the entire expression, radical sign and radicand, is called a **radical**. An algebraic expression containing a radical is called a **radical expression**.

Definition: Square Roots of a

If a is positive real number,

$$\sqrt{a} \text{ is the positive square root of } a,$$
$$-\sqrt{a} \text{ is the negative square root of } a.$$

For nonnegative a,

$$\sqrt{a} \cdot \sqrt{a} = \left(\sqrt{a}\right)^2 = a \quad \text{and} \quad -\sqrt{a} \cdot -\sqrt{a} = \left(-\sqrt{a}\right)^2 = a.$$

Also, $\sqrt{0} = 0$.

EXAMPLE 1
Find each square root.

(a) $\sqrt{169}$

The radical $\sqrt{169}$ represents the positive square root of 169. Think of a positive number whose square is 169.

$$13^2 = 169, \text{ so } \sqrt{169} = 13.$$

(b) $-\sqrt{10,000}$

This symbol represents the negative square root of 10,000. A calculator with a square root key can be used to find $\sqrt{10,000} = 100$. Then, $-\sqrt{10,000} = -100$.

When the square root of a positive real number is squared, the result is that positive real number. (Also $(\sqrt{0})^2 = 0$.)

EXAMPLE 2

(a) $\left(\sqrt{11}\right)^2 = 11$ Definition of square root

(b) $\left(-\sqrt{31}\right)^2 = 31$ The square of a *negative* number is positive.

Perfect Square: All numbers with square roots that are rational are called **perfect squares**. For example, 100 and $\frac{25}{36}$ are perfect squares. A number that is not a perfect square has a square root that is not a rational number. For example, $\sqrt{10}$ is not a rational number because it cannot be written as the ratio of two integers. We can find a decimal approximation for $\sqrt{10}$ by using a calculator. For example, a calculator shows that $\sqrt{10}$ is 3.16227766, although this is only a rational number *approximation* of $\sqrt{10}$.

EXAMPLE 3
Find a decimal approximation for each square root. Round answers to the nearest thousandth.

(a) $\sqrt{12}$

Using the square root key of a calculator gives $\sqrt{12} \approx 3.46410162 \approx 3.464$ rounded to the nearest thousandth, where $\approx$ means "is approximately equal to."

(b) $\sqrt{27} \approx 5.196$

(c) $-\sqrt{10,134} \approx -100.668$

Self-Check 1

Find each of the following.

1. $-\sqrt{256}$ 2. $\left(\sqrt{81}\right)^2$

Find a decimal approximation for each square root. Round answers to the nearest thousandth.

3. $-\sqrt{127}$ 4. $\sqrt{51796}$

The Pythagorean formula: One application of square roots uses the Pythagorean formula. Recall from geometry that by this formula, if c is the length of the hypotenuse of a right triangle, and a and b are the lengths of the two legs, then

$$c^2 = a^2 + b^2.$$

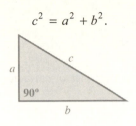

EXAMPLE 4

Find the unknown length of the third side of the right triangle with sides $a = 5$ and $b = 12$, where c is the hypotenuse.

Use the formula to find c^2 first.

$$
\begin{aligned}
c^2 &= a^2 + b^2 \\
&= 5^2 + 12^2 &&\text{Let } a = 5 \text{ and } b = 12. \\
&= 25 + 144 = 169 &&\text{Square and add.}
\end{aligned}
$$

Now find the positive square root of 169 to get c.

$$c = \sqrt{169} = 13$$

(Although -13 is also a square root of 169, the length of a side of a triangle must be a positive number.)

Be careful not to make the common mistake of thinking that $\sqrt{a^2 + b^2}$ equals $a + b$. As Example 4 shows,

$$\sqrt{25 + 144} = \sqrt{169} = 13 \neq \sqrt{25} + \sqrt{144} = 5 + 12 = 17$$

so that, in general,

$$\sqrt{a^2 + b^2} \neq a + b.$$

EXAMPLE 5

A ladder 15 feet long leans against a wall. The foot of the ladder is 9 feet from the base of the wall. How high up the wall does the top of the ladder rest?

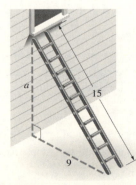

As shown in the diagram, a right triangle is formed with the ladder as the hypotenuse. Let a represent the height of the top of the ladder. By the Pythagorean formula,

$$c^2 = a^2 + b^2.$$
$$15^2 = a^2 + 9^2 \qquad \text{Let } c = 15 \text{ and } b = 9.$$
$$225 = a^2 + 81 \qquad \text{Square.}$$
$$144 = a^2 \qquad \text{Subtract 81.}$$
$$\sqrt{144} = a$$
$$a = 12 \qquad\qquad \sqrt{144} = 12$$

Choose the positive square root of 144 since a represents a length.

Check.

$$15^2 = 12^2 + 9^2$$
$$225 = 144 + 81$$
$$225 = 225$$

The top of the ladder rests 12 feet up the wall.

Self-Check 2

1. Find the unknown length of the third side (to the nearest thousandths) of the right triangle with sides $a = 7$ and $b = 10$, where c is the hypotenuse.

2. What is the base of a right triangle if it's height is 24 cm and the hypotenuse is 26 cm?

Higher roots; Principal roots: Finding the square root of a number is the inverse (reverse) of squaring a number. In a similar way, there are inverses to finding the cube of a number, or finding the fourth power of a number. These inverses are the **cube root**, written $\sqrt[3]{a}$ and the **fourth root**, written $\sqrt[4]{a}$. Similar symbols are used for higher roots. In general we the nth root of the number a is written $\sqrt[n]{a}$.

In $\sqrt[n]{a}$, the number n is the **index** or **order** of the radical. A calculator that has a key marked $\boxed{\sqrt[x]{y}}$ or $\boxed{x^y}$ can be used to find these roots. When working with cube roots or fourth roots, it is helpful or to memorize the first few *perfect cubes* ($2^3 = 8$, $3^3 = 27$, and so on) and the first few perfect fourth powers.

EXAMPLE 6
Find each cube root.

(a) $\sqrt[3]{64}$

Look for a number that can be cubed to give 64. Since $4^3 = 64$, then $\sqrt[3]{64} = 4$.

(b) $\sqrt[3]{-64} = -4$ because $(-4)^3 = -64$.

As Example 6 suggests, the cube root of a positive number is positive, and the cube root of a negative number is negative. *There is only one real number cube root for each real number.*

When the index of the radical is even (square root, fourth root, and so on), the radicand must be nonnegative to get a real number root. Also, for even indexes the symbols $\sqrt{}$, $\sqrt[4]{}$, $\sqrt[6]{}$, and so on are used for the *nonnegative* roots, which are called **principal roots**. The symbols $-\sqrt{}$, $-\sqrt[4]{}$, $-\sqrt[6]{}$ and so on are used for the negative roots.

EXAMPLE 7
Find each root.

(a) $\sqrt[4]{81} = 3$ because 2 is positive and $3^4 = 81$.

(b) $-\sqrt[4]{81} = -3$ because $\sqrt[4]{81} = 3$ from part (a).

(c) $\sqrt[4]{-81}$
To find the fourth root, the radicand must be nonnegative. There is no real number that equals $\sqrt[4]{-81}$.

(d) $\sqrt[5]{32} = 2$ since $2^5 = 32$.

(e) $\sqrt[5]{-32} = -2$ since $(-2)^5 = -32$.

Self-Check 3

Find each root.

1. $\sqrt[3]{125}$ 2. $\sqrt[3]{-1000}$ 3. $-\sqrt[3]{1000}$ 4. $-\sqrt[6]{64}$

Self-Check Answers

1.1 −16	**1.2** 81	**1.3** −11.269	**1.4** 227.587
2.1 20.207		**2.2** 10cm	
3.1 5	**3.2** −10	**3.3** −10	**3.4** −2

5.1 EXERCISES

In exercises 1-6, decide whether each statement is true or false. If false, tell why.

1. Every nonnegative number has two square roots.

2. A negative number has negative square roots.

3. Every positive number has two real square roots.

4. Every positive number has three real cube roots.

5. The cube root of every real number has the same sign as the number itself.

6. The positive square root of a positive number is its principal square root.

Find all square roots of each number in exercises 7-14.

7. 4   8. 1   9. 196   10. 289

11. $\dfrac{4}{25}$ 12. $\dfrac{81}{100}$ 13. 2500   14. 3600

Find each square root that is a real number in exercises 15-22.

15. $\sqrt{36}$ 16. $\sqrt{121}$ 17. $-\sqrt{361}$ 18. $-\sqrt{441}$

19. $-\sqrt{\dfrac{4}{9}}$ 20. $-\sqrt{\dfrac{25}{169}}$ 21. $\sqrt{-49}$ 22. $\sqrt{-81}$

Use a calculator with a square root key to find each root in exercises 23-28. Round to the nearest thousandth.

23. $\sqrt{123}$ **24.** $\sqrt{456}$ **25.** $\sqrt{7.1549}$

26. $\sqrt{3.14}$ **27.** $\sqrt{126.63}$ **28.** $\sqrt{0.042756}$

Find each square root in exercises 29-34. Use a calculator and round to the nearest thousandth, if necessary. (Hint: First simplify the radicand to a single number.)

29. $\sqrt{18^2 + 24^2}$ **30.** $\sqrt{15^2 + 20^2}$ **31.** $\sqrt{16^2 + (-30)^2}$

32. $\sqrt{(-15)^2 + 36^2}$ **33.** $\sqrt{5^2 + 6^2}$ **34.** $\sqrt{7^2 + 8^2}$

Use a calculator with a cube root key to find each root in exercises 35-38. Round to the nearest thousandth.

35. $\sqrt[3]{9}$ **36.** $\sqrt[3]{25}$ **37.** $\sqrt[3]{96.1}$ **38.** $\sqrt[3]{102.9}$

For exercises 39-44, find the length of the unknown side of each right triangle with legs a and b and hypotenuse c. In exercises 43 and 44, use a calculator and round to the nearest thousandth.

39. $a = 24, b = 32$ **40.** $a = 36, b = 15$ **41.** $a = 8, c = 10$

42. $b = 5, c = 13$ **43.** $a = 7, b = 9$ **44.** $a = 19, b = 11$

Use the Pythagorean formula to solve each problem in exercises 45-48.

45. The diagonal of a rectangle measures 41 centimeters. The width of the rectangle is 9 centimeters. Find the length of the rectangle.

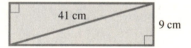

46. The length of a rectangle is 24 meters and the width is 7 meters. Find the measure of the diagonal of the rectangle.

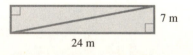

47. Margaret is flying a kite on 125 feet of string. How high is it above her hand (vertically) if the horizontal distance between Margaret and the kite is 75 feet?

48. A guy wire is attached to the mast of a short-wave transmitting antenna. It is attached 104 feet above ground level. If the wire is staked to the ground 78 feet from the base of the mast, then how long is the wire?

49. The figure on the left illustrates the Pythagorean formula by using a tile pattern. In the figure, the side of the square along the hypotenuse measures 5 units, while the sides along the legs measure 3 and 4 units. If we let $a = 3$, $b = 4$, and $c = 5$, the equation of the Pythagorean formula is satisfied.

$$a^2 + b^2 \overset{?}{=} c^2$$
$$3^2 + 4^2 \overset{?}{=} 5^2$$
$$9 + 16 \overset{?}{=} 25$$
$$25 = 25 \quad \text{True}$$

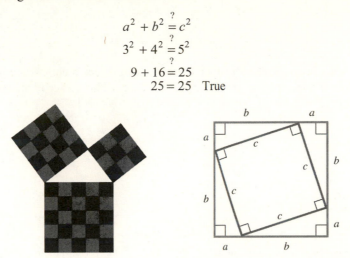

Use the diagram on the right to verify the Pythagorean formula. (To do so, express the area of the figure in two ways: first, as the area of the large square, and then as the sum of the areas of the smaller square and the four right triangles. Finally, set the areas equal and simplify the equation.)

5.2 | Multiplication and Division of Radicals

Multiply Radicals: The product rule for radicals is important in multiplication of radicals. To illustrate the rule, notice that

$$\sqrt{4} \cdot \sqrt{25} = 2 \cdot 5 = 10 \quad \text{and} \quad \sqrt{4 \cdot 25} = \sqrt{100} = 10,$$

showing that

$$\sqrt{4} \cdot \sqrt{25} = \sqrt{4 \cdot 25}.$$

This result is a particular case of the more general product rule for radicals.

Product Rule for Radicals

For nonnegative real numbers x and y,

$$\sqrt{x} \cdot \sqrt{y} = \sqrt{x \cdot y} \quad \text{and} \quad \sqrt{x \cdot y} = \sqrt{x} \cdot \sqrt{y}$$

That is, the product of two radicals is the radical of the product.

In general, $\sqrt{x + y} \neq \sqrt{x} + \sqrt{y}$. To see why this is so, let $x = 81$ and $y = 144$.

$$\sqrt{81 + 144} = \sqrt{225} = 15$$

but

$$\sqrt{81} + \sqrt{144} = 9 + 12 = 21.$$

EXAMPLE 1

Use the product rule for radicals to find each product.

(a) $\sqrt{3} \cdot \sqrt{5} = \sqrt{3 \cdot 5} = \sqrt{15}$

(b) $\sqrt{11} \cdot \sqrt{7} = \sqrt{11 \cdot 7} = \sqrt{77}$

(c) $\sqrt{2} \cdot \sqrt{a} = \sqrt{2a}$ Assume $a \geq 0$.

Simplify Radicals Using the Product Rule: A square root radical is **simplified** when no perfect square factor remains under the radical sign. This is accomplished by using the product rule as shown in Example 2.

EXAMPLE 2

Simplify each radical.

(a) $\sqrt{24} = \sqrt{4 \cdot 6}$ 4 is a perfect square.

$\quad\quad = \sqrt{4} \cdot \sqrt{6}$ Product Rule

$\quad\quad = 2\sqrt{6}$ $\sqrt{4} = 2$

Since 6 has no perfect square factor other than 1, $2\sqrt{6}$ is called the **simplified form** of $\sqrt{24}$.

(b) $\sqrt{108} = \sqrt{36 \cdot 3}$ 36 is the largest perfect square factor of 108.

$\quad\quad = \sqrt{36} \cdot \sqrt{3}$ Product Rule

$\quad\quad = 6\sqrt{3}$ $\sqrt{36} = 6$

(c) $\sqrt{10}$

The number 10 has no perfect square factor (except 1), so $\sqrt{10}$ cannot be simplified further.

EXAMPLE 3

Find the product and simplify.

$\sqrt{16} \cdot \sqrt{50} = 4\sqrt{50}$ $\sqrt{16} = 4$

$\quad\quad = 4\sqrt{25 \cdot 2}$ 25 is a perfect square.

$\quad\quad = 4\sqrt{25} \cdot \sqrt{2}$ Product Rule

$\quad\quad = 4 \cdot 5\sqrt{2}$ $\sqrt{25} = 5$

$\quad\quad = 20\sqrt{2}$ Multiply.

Self-Check 1

Find each product and simplify, if possible.

1. $\sqrt{6} \cdot \sqrt{7}$ 　　　　　　　　　**2.** $\sqrt{6} \cdot \sqrt{12}$

3. $\sqrt{16} \cdot \sqrt{27}$ 　　　　　　　　**4.** $\sqrt{12} \cdot \sqrt{72}$

Simplify Radicals Using the Quotient Rule: The *quotient rule for radicals* is very similar to the product rule.

Quotient Rule for Radicals

For nonnegative real numbers x and $y \neq 0$,

$$\sqrt{\frac{x}{y}} = \frac{\sqrt{x}}{\sqrt{y}} \quad \text{and} \quad \frac{\sqrt{x}}{\sqrt{y}} = \sqrt{\frac{x}{y}}.$$

That is, the radical of the quotient is the quotient of the radicals.

EXAMPLE 4
Simplify each radical.

(a) $\sqrt{\frac{36}{49}} = \frac{\sqrt{36}}{\sqrt{49}} = \frac{6}{7}$ Quotient Rule

(b) $\frac{\sqrt{845}}{\sqrt{5}} = \sqrt{\frac{845}{5}} = \sqrt{169} = 13$ Quotient Rule

(c) $\sqrt{\frac{61}{9}} = \frac{\sqrt{61}}{\sqrt{9}} = \frac{\sqrt{61}}{3}$ Quotient Rule

EXAMPLE 5

Divide $24\sqrt{35}$ by $6\sqrt{7}$.

We use the quotient rule as follows.

$$\frac{24\sqrt{35}}{6\sqrt{7}} = \frac{24}{6} \cdot \frac{\sqrt{35}}{\sqrt{7}} = 4\sqrt{\frac{35}{7}} = 4\sqrt{5}$$

Some problems require both the product and quotient rules, as Example 6 shows.

EXAMPLE 6

Simplify $\sqrt{\frac{7}{11}} \cdot \sqrt{\frac{9}{11}}$.

$$\sqrt{\frac{7}{11}} \cdot \sqrt{\frac{9}{11}} = \frac{\sqrt{7}}{\sqrt{11}} \cdot \frac{\sqrt{9}}{\sqrt{11}} \qquad \text{Quotient Rule}$$

$$= \frac{\sqrt{7}\sqrt{9}}{\sqrt{11}\sqrt{11}} \qquad \text{Multiply fractions.}$$

$$= \frac{\sqrt{7} \cdot 3}{\sqrt{121}} \qquad \text{Product rule; } \sqrt{9} = 3$$

$$= \frac{3\sqrt{7}}{11} \qquad \sqrt{121} = 11$$

Self-Check 2

Simplify.

1. $\sqrt{\frac{81}{25}}$ 2. $\sqrt{\frac{97}{16}}$

Perform the indicated operation and simplify.

3. Divide $28\sqrt{143}$ by $7\sqrt{11}$ 4. $\sqrt{\frac{17}{13}} \cdot \sqrt{\frac{16}{13}}$

The product and quotient rules also apply when variables appear under the radical sign, as long as all the variables represent only nonnegative numbers. For example, $\sqrt{6^2} = 6$, but $\sqrt{(-6)^2} \neq -6$.

For a real number a, $\sqrt{a^2} = a$ only if a is nonnegative. If we assume, for example, that m is nonnegative, then $\sqrt{36m^6} = 6m^3$.

Simplify Higher Roots: To simplify cube roots, look for factors that are *perfect cubes*. A **perfect cube** is a number with a rational cube root. For example, $\sqrt[3]{27} = 3$, and since 3 is a rational number, 27 is a perfect cube. Higher roots are handled in a similar manner.

Properties of Radicals

For all real numbers where the indicated roots exist,

$$\sqrt[n]{x} \cdot \sqrt[n]{y} = \sqrt[n]{xy} \quad \text{and} \quad \frac{\sqrt[n]{x}}{\sqrt[n]{y}} = \sqrt[n]{\frac{x}{y}}, \ y \neq 0.$$

EXAMPLE 7
Simplify each radical.

(a) $\sqrt[3]{243} = \sqrt[3]{27 \cdot 9}$ 27 is a perfect cube.

$\qquad\quad = \sqrt[3]{27} \cdot \sqrt[3]{9} = 3\sqrt[3]{9}$

(b) $\sqrt[4]{243} = \sqrt[4]{81 \cdot 3}$ 81 is a perfect fourth power.

$\qquad\quad = \sqrt[4]{81} \cdot \sqrt[4]{3} = 3\sqrt[4]{3}$

(c) $\sqrt[3]{\dfrac{27}{125}} = \dfrac{\sqrt[3]{27}}{\sqrt[3]{125}} = \dfrac{3}{5}$

Self-Check 3

Simplify.

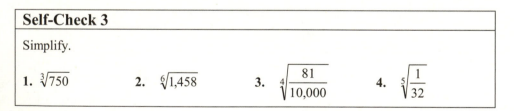

1. $\sqrt[3]{750}$ 2. $\sqrt[6]{1,458}$ 3. $\sqrt[4]{\dfrac{81}{10,000}}$ 4. $\sqrt[5]{\dfrac{1}{32}}$

Rationalize Denominators: Fractions are simplified by rewriting them without any radical expressions in the denominator. For example, the radical in the denominator of

$$\frac{\sqrt{5}}{\sqrt{3}}$$

can be eliminated by multiplying the numerator and the denominator by $\sqrt{3}$.

This process of changing the denominator from a radical (irrational number) to a rational number is called **rationalizing the denominator**. The value of the number is not changed; only the form of the number is changed, because the expression has been multiplied by 1 in the form $\dfrac{\sqrt{3}}{\sqrt{3}}$.

EXAMPLE 8

Rationalize each denominator.

(a) $\dfrac{35}{\sqrt{10}} = \dfrac{35 \cdot \sqrt{10}}{\sqrt{10} \cdot \sqrt{10}}$ Multiply by $\dfrac{\sqrt{10}}{\sqrt{10}}$.

$\phantom{\dfrac{35}{\sqrt{10}}} = \dfrac{35\sqrt{10}}{10}$ $\sqrt{10} \cdot \sqrt{10} = 10$

$\phantom{\dfrac{35}{\sqrt{10}}} = \dfrac{7\sqrt{10}}{2}$

(b) $\dfrac{30}{\sqrt{12}}$

The denominator here could be rationalized by multiplying by $\sqrt{12}$. However, the result can be found more directly by first simplifying the denominator.

$$\sqrt{12} = \sqrt{4 \cdot 3} = \sqrt{4} \cdot \sqrt{3} = 2\sqrt{3}$$

Then multiply numerator and denominator by $\sqrt{3}$.

$$\dfrac{30}{\sqrt{12}} = \dfrac{30}{2\sqrt{3}}$$

$$\phantom{\dfrac{30}{\sqrt{12}}} = \dfrac{15}{\sqrt{3}}$$

$$\phantom{\dfrac{30}{\sqrt{12}}} = \dfrac{15 \cdot \sqrt{3}}{\sqrt{3} \cdot \sqrt{3}} \qquad \text{Multiply by } \dfrac{\sqrt{3}}{\sqrt{3}}.$$

$$\phantom{\dfrac{30}{\sqrt{12}}} = \dfrac{15\sqrt{3}}{3} \qquad \sqrt{3} \cdot \sqrt{3} = 3$$

$$\phantom{\dfrac{30}{\sqrt{12}}} = 5\sqrt{3} \qquad \text{Lowest terms}$$

Self-Check 4

Rationalize each denominator.

1. $\dfrac{9}{\sqrt{7}}$ 2. $\dfrac{6}{\sqrt{20}}$

Simplified Form: A radical is considered to be in simplified form if the following three conditions are met.

Definition: Simplified Form of a Radical

1. All nth power factors of the radicand $\sqrt[n]{}$ of are removed. (For example, $\sqrt{5^2} = 5$, $\sqrt[3]{7^3} = 7$, and so on.)

2. The radicand has no fractions.

3. No denominator contains a radical.

In the following examples, radicals are put in simplified form. We show simplifying a fraction, a product, and a quotient.

EXAMPLE 9

(a) Simplify $\sqrt{\dfrac{128}{7}}$ by rationalizing the denominator.

First use the quotient rule for radicals. Then multiply both numerator and denominator by $\sqrt{7}$.

$$\sqrt{\frac{128}{7}} = \frac{\sqrt{128}\cdot\sqrt{7}}{\sqrt{7}\cdot\sqrt{7}} \qquad \text{Rationalize the denominator.}$$

$$= \frac{\sqrt{64\cdot 2}\cdot\sqrt{7}}{7} \qquad \sqrt{7}\cdot\sqrt{7}=7$$

$$= \frac{\sqrt{64}\cdot\sqrt{2}\cdot\sqrt{7}}{7} \qquad \text{Product rule}$$

$$= \frac{8\sqrt{2\cdot 7}}{7} = \frac{8\sqrt{14}}{7} \qquad \text{Product rule}$$

(b) $\sqrt{\dfrac{7}{12}}\cdot\sqrt{\dfrac{1}{15}} = \sqrt{\dfrac{7}{12}\cdot\dfrac{1}{15}} \qquad \text{Product Rule}$

$$= \sqrt{\frac{7}{180}} \qquad \text{Multiply}$$

$$= \frac{\sqrt{7}}{\sqrt{180}} \qquad \text{Quotient Rule}$$

To rationalize the denominator, first simplify, then multiply the numerator and denominator by $\sqrt{5}$ as follows.

$$\frac{\sqrt{7}}{\sqrt{180}} = \frac{\sqrt{7}}{\sqrt{36}\sqrt{5}} \qquad \text{Product Rule}$$

$$= \frac{\sqrt{7}}{6\sqrt{5}} \qquad \sqrt{36}=6$$

$$= \frac{\sqrt{7}\cdot\sqrt{5}}{6\sqrt{5}\cdot\sqrt{5}} \qquad \text{Rationalize the denominator.}$$

$$= \frac{\sqrt{35}}{6\cdot 5} \qquad \text{Product Rule}$$

$$= \frac{\sqrt{35}}{30}$$

(c) Rationalize the denominator of $\sqrt{\dfrac{9a}{b}}$. Assume that a and b represent positive real numbers.

Multiply the numerator and denominator by $\sqrt{b}$.

$$\frac{\sqrt{9a}}{\sqrt{b}} = \frac{\sqrt{9a}\cdot\sqrt{b}}{\sqrt{b}\cdot\sqrt{b}} = \frac{\sqrt{9ab}}{b} = \frac{3\sqrt{ab}}{b}$$

Self-Check 5

Simplify. Assume that all variables represent positive real numbers.

1. $\sqrt{\dfrac{300}{11}}$ 2. $\sqrt{\dfrac{1}{20}}\cdot\sqrt{\dfrac{3}{55}}$ 3. $\dfrac{24\sqrt{3}}{\sqrt{6}}$ 4. $\sqrt{\dfrac{12x}{y}}$

Self-Check Answers

1.1 $\sqrt{42}$	**1.2** $6\sqrt{2}$	**1.3** $12\sqrt{3}$	**1.4** $12\sqrt{6}$
2.1 $\dfrac{9}{5}$	**2.2** $\dfrac{\sqrt{97}}{4}$	**2.3** $4\sqrt{13}$	**2.4** $\dfrac{4\sqrt{17}}{13}$
3.1 $5\sqrt[3]{6}$	**3.2** $3\sqrt[6]{2}$	**3.3** $\dfrac{3}{10}$	**3.4** $\dfrac{1}{2}$
4.1 $\dfrac{9\sqrt{7}}{7}$		**4.2** $\dfrac{3\sqrt{5}}{5}$	
5.1 $\dfrac{10\sqrt{33}}{11}$	**5.2** $\dfrac{\sqrt{33}}{110}$	**5.3** $12\sqrt{2}$	**5.4** $\dfrac{2\sqrt{3xy}}{y}$

5.2 EXERCISES

Use the product rule for radicals to find each product in exercises 1-8.

1. $\sqrt{2} \cdot \sqrt{8}$ 2. $\sqrt{5} \cdot \sqrt{125}$ 3. $\sqrt{6} \cdot \sqrt{12}$ 4. $\sqrt{10} \cdot \sqrt{20}$

5. $\sqrt{21} \cdot \sqrt{21}$ 6. $\sqrt{29} \cdot \sqrt{29}$ 7. $\sqrt{6} \cdot \sqrt{x}, x \geq 0$ 8. $\sqrt{5} \cdot \sqrt{y}, y \geq 0$

Simplify each radical in exercises 9-18.

9. $\sqrt{63}$ 10. $\sqrt{84}$

11. $\sqrt{108}$ 12. $\sqrt{162}$

13. $\sqrt{200}$ 14. $\sqrt{300}$

15. $-\sqrt{800}$ 16. $-\sqrt{1200}$

17. $13\sqrt{12}$ 18. $10\sqrt{80}$

Find each product and simplify in exercises 19-24.

19. $\sqrt{2} \cdot \sqrt{14}$ 20. $\sqrt{3} \cdot \sqrt{15}$

21. $\sqrt{6} \cdot \sqrt{24}$ 22. $\sqrt{50} \cdot \sqrt{8}$

23. $\sqrt{14} \cdot \sqrt{21}$ 24. $\sqrt{18} \cdot \sqrt{24}$

25. In your own words, describe the product and quotient rules.

26. Simplify the radical $\sqrt{432}$ in two ways. First, factor 432 as $144 \cdot 3$ and then simplify completely. Second, factor 432 as $36 \cdot 12$ and then simplify completely. How do the answers compare? Make a conjecture concerning the quickest way to simplify such a radical.

Use the quotient rule and the product rule, as necessary, to simplify each radical expression in exercises 27-32.

27. $\sqrt{\dfrac{25}{196}}$ 28. $\sqrt{\dfrac{36}{529}}$

29. $\sqrt{\dfrac{2}{11}} \cdot \sqrt{22}$

30. $\sqrt{\dfrac{3}{13}} \cdot \sqrt{39}$

31. $\dfrac{24\sqrt{20}}{8\sqrt{5}}$

32. $\dfrac{42\sqrt{39}}{7\sqrt{3}}$

Simplify each radical in exercises 33-38. Assume that all variables represent nonnegative real numbers.

33. $\sqrt{z^2}$

34. $\sqrt{c^2}$

35. $\sqrt{f^6}$

36. $\sqrt{g^8}$

37. $\sqrt{400x^{10}}$

38. $\sqrt{1600b^{12}}$

Simplify each radical in exercises 39-46.

39. $\sqrt[3]{56}$

40. $\sqrt[3]{72}$

41. $\sqrt[3]{135}$

42. $\sqrt[3]{162}$

43. $\sqrt[4]{96}$

44. $\sqrt[4]{405}$

45. $\sqrt[3]{\dfrac{27}{64}}$

46. $\sqrt[3]{\dfrac{8}{125}}$

Fill in each blank with the correct response in exercises 47-50.

47. Rationalizing the denominator means to change the denominator from a(n) _____ to a rational number.

48. To simplify $\sqrt{a^2b}$, where $a,b \geq 0$, we must remove _____ from the radicand.

49. The expression $\sqrt{\dfrac{3z}{5}}$, where $z \geq 0$, is not simplified because the radical contains a(n) _____.

50. The expression $\dfrac{2s}{\sqrt{10}}$ is not simplified because the denominator is a(n) _____.

Rationalize each denominator in exercises 51-66.

51. $\dfrac{10}{\sqrt{5}}$

52. $\dfrac{15}{\sqrt{3}}$

53. $-\dfrac{\sqrt{13}}{\sqrt{5}}$

54. $-\dfrac{\sqrt{23}}{\sqrt{7}}$

55. $\dfrac{10\sqrt{10}}{5\sqrt{3}}$

56. $\dfrac{8\sqrt{13}}{4\sqrt{7}}$

57. $\dfrac{10}{\sqrt{27}}$

58. $\dfrac{7}{\sqrt{50}}$

59. $\dfrac{63}{\sqrt{18}}$

60. $\dfrac{52}{\sqrt{20}}$

61. $\dfrac{\sqrt{40}}{\sqrt{20}}$

62. $\dfrac{\sqrt{96}}{\sqrt{32}}$

63. $\sqrt{\dfrac{1}{2}}$

64. $\sqrt{\dfrac{1}{3}}$

65. $\sqrt{\dfrac{5}{7}}$

66. $\sqrt{\dfrac{7}{11}}$

Multiply and simplify each result in exercises 67-74.

67. $\sqrt{\dfrac{11}{17}} \cdot \sqrt{\dfrac{17}{5}}$ **68.** $\sqrt{\dfrac{17}{21}} \cdot \sqrt{\dfrac{21}{2}}$ **69.** $\sqrt{\dfrac{17}{7}} \cdot \sqrt{\dfrac{17}{27}}$ **70.** $\sqrt{\dfrac{7}{6}} \cdot \sqrt{\dfrac{7}{12}}$

71. $\sqrt{\dfrac{1}{18}} \cdot \sqrt{\dfrac{1}{2}}$ **72.** $\sqrt{\dfrac{1}{20}} \cdot \sqrt{\dfrac{1}{5}}$ **73.** $\sqrt{\dfrac{3}{4}} \cdot \sqrt{\dfrac{4}{3}}$ **74.** $\sqrt{\dfrac{21}{29}} \cdot \sqrt{\dfrac{29}{21}}$

Simplify each radical in exercises 75-82. Assume that all variables represent positive real numbers.

75. $\sqrt{\dfrac{2}{x}}$ **76.** $\sqrt{\dfrac{17}{y}}$ **77.** $\sqrt{\dfrac{4s^3}{t}}$ **78.** $\sqrt{\dfrac{14a^3}{b}}$

79. $\sqrt{\dfrac{24x^3}{8y}}$ **80.** $\sqrt{\dfrac{10m^3}{2n}}$ **81.** $\sqrt{\dfrac{25a^2b^5}{3c}}$ **82.** $\sqrt{\dfrac{49x^3y^2}{11z}}$

5.3   Addition and Subtraction of Radicals

Like radicals: We add or subtract radicals by using the distributive property. For example,

$$7\sqrt{2} + 5\sqrt{2} = (7 + 5)\sqrt{2} \quad \text{Distributive property}$$
$$= 12\sqrt{2}$$

Also,

$$2\sqrt{13} - 8\sqrt{13} = -6\sqrt{13}.$$

Like radicals are terms that have multiples of the *same root* of the *same number*. Only like radicals can be combined using the distributive property. In the example above, the like radicals are $2\sqrt{13}$ and $-8\sqrt{13}$. On the other hand, examples of *unlike radicals* are

$$4\sqrt{3} \quad \text{and} \quad 4\sqrt{7}, \quad \text{Different radicals}$$

as well as

$$3\sqrt{5} \quad \text{and} \quad 3\sqrt[3]{5}. \quad \text{Different indexes}$$

EXAMPLE 1

(a) $3\sqrt{10} + 6\sqrt{10} = (3 + 6)\sqrt{10} = 9\sqrt{10}$ Distributive property

(b) $3\sqrt{10} - 6\sqrt{10} = (3 - 6)\sqrt{10} = -3\sqrt{10}$

(c) $\sqrt[3]{4} + \sqrt[3]{4} = 1 \cdot \sqrt[3]{4} + 1 \cdot \sqrt[3]{4} = (1 + 1)\sqrt[3]{4} = 2\sqrt[3]{4}$

(d) $\sqrt[5]{8} + 3\sqrt[5]{8} = 1 \cdot \sqrt[5]{8} + 3\sqrt[5]{8} = (1 + 3)\sqrt[5]{8} = 4\sqrt[5]{8}$

(e) $\sqrt{2} + \sqrt{3}$ cannot be added using the distributive property.

Self-Check 1

Add or subtract as indicated.

1. $6\sqrt{6} + 4\sqrt{6}$ $\qquad\qquad$ **2.** $12\sqrt[3]{5} - 2\sqrt[3]{5}$

3. $\sqrt[5]{17} + 8\sqrt[5]{17}$ $\qquad\qquad$ **4.** $4\sqrt{2} + 3\sqrt[3]{2}$

Simplify Radicals; Add and Subtract: Sometimes we must simplify one or more radicals in a sum or difference. When this results in like radicals, we can then add or subtract.

EXAMPLE 2
Simplify as much as possible.

(a) $3\sqrt{2} + \sqrt{50} = 3\sqrt{2} + \sqrt{25 \cdot 2}$ $\qquad$ Factor

$\qquad\qquad\qquad = 3\sqrt{2} + \sqrt{25} \cdot \sqrt{2}$ $\qquad$ Product Rule

$\qquad\qquad\qquad = 3\sqrt{2} + 5\sqrt{2}$ $\qquad\qquad$ $\sqrt{25} = 5$

$\qquad\qquad\qquad = 8\sqrt{2}$ $\qquad\qquad\qquad$ Add like radicals.

(b) $5\sqrt[3]{54} + 3\sqrt[3]{2} = 5\left(\sqrt[3]{27} \cdot \sqrt[3]{2}\right) + 3\sqrt[3]{2}$ $\qquad$ Product rule

$\qquad\qquad\qquad = 5\left(3\sqrt[3]{2}\right) + 3\sqrt[3]{2}$ $\qquad\qquad$ $\sqrt[3]{27} = 3$

$\qquad\qquad\qquad = 15\sqrt[3]{2} + 3\sqrt[3]{2}$ $\qquad\qquad$ Multiply.

$\qquad\qquad\qquad = 18\sqrt[3]{2}$ $\qquad\qquad\qquad$ Add like radicals.

Self-Check 2

Simplify as much as possible.

1. $5\sqrt{12} + 6\sqrt{3}$ $\qquad\qquad$ **2.** $4\sqrt{80} - 7\sqrt{45}$

3. $3\sqrt[3]{500} + 4\sqrt[3]{108}$ $\qquad\qquad$ **4.** $\sqrt[5]{96} + \sqrt[5]{128}$

Simplify Radicals; Multiply: Some radical expressions require both multiplication and addition (or subtraction). The order of operations presented in Chapter 1 still applies.

EXAMPLE 3
Simplify each radical expression. Assume that all variables represent nonnegative real numbers.

(a) $\sqrt{5} \cdot \sqrt{10x} + 7\sqrt{8x} = \sqrt{50x} + 7\sqrt{8x}$ $\qquad\qquad$ Product rule

$\qquad\qquad\qquad\qquad = \sqrt{25 \cdot 2x} + 7\sqrt{4 \cdot 2x}$ $\qquad\qquad$ Factor.

$\qquad\qquad\qquad\qquad = \sqrt{25} \cdot \sqrt{2x} + 7\sqrt{4} \cdot \sqrt{2x}$ $\qquad$ Product rule

$\qquad\qquad\qquad\qquad = 5\sqrt{2x} + 7 \cdot 2\sqrt{2x}$ $\qquad\qquad$ $\sqrt{25} = 5$ and $\sqrt{4} = 2$

$\qquad\qquad\qquad\qquad = 5\sqrt{2x} + 14\sqrt{2x}$ $\qquad\qquad$ Multiply

$\qquad\qquad\qquad\qquad = 19\sqrt{2x}$ $\qquad\qquad\qquad$ Add like radicals.

(b) $\sqrt[3]{3} \cdot \sqrt[3]{18a^3} - 5\sqrt[3]{16a^3} = \sqrt[3]{54a^3} - 5\sqrt[3]{16a^3}$ Product rule

$\qquad = \sqrt[3]{(27a^3) \cdot 2} - 5\sqrt[3]{(8a^3) \cdot 2}$ Factor.

$\qquad = 3a\sqrt[3]{2} - 5 \cdot 2a\sqrt[3]{2}$ $\sqrt[3]{27a^3} = 3a$ and $\sqrt[3]{8a^3} = 2a$

$\qquad = 3a\sqrt[3]{2} - 10a\sqrt[3]{2}$ Multiply

$\qquad = -7a\sqrt[3]{2}$ Subtract like radicals.

Remember that a sum or difference of radicals can be simplified only if the radicals are *like radicals*. For example, $11\sqrt{2} + 6\sqrt[3]{2}$ cannot be simplified further.

Self-Check 3

Simplify each radical expression. Assume that all variables represent nonnegative real numbers.

1. $\sqrt{3} \cdot \sqrt{15d} + 8\sqrt{20d}$ 2. $\sqrt{24f} \cdot \sqrt{14g} - 2\sqrt{84fg}$

3. $\sqrt[3]{2} \cdot \sqrt[3]{12y^3} + 6\sqrt[3]{81y^3}$ 4. $\sqrt[3]{3t} \cdot \sqrt[3]{18s^6} - 6s^2\sqrt[3]{2t}$

Self-Check Answers

1.1 $10\sqrt{6}$ 1.2 $10\sqrt[3]{5}$

1.3 $9\sqrt[5]{17}$ 1.4 cannot be simplified further

2.1 $16\sqrt{3}$ 2.2 $-5\sqrt{5}$ 2.3 $27\sqrt[3]{4}$ 2.4 $2\sqrt[5]{3} + 2\sqrt[5]{4}$

3.1 $19\sqrt{5d}$ 3.2 0   3.3 $20y\sqrt[3]{3}$ 3.4 $-3s^2\sqrt[3]{2t}$

5.3 EXERCISES

Fill in each blank with the correct response in exercises 1-4.

1. $6\sqrt{5} + 2\sqrt{5} = (6 + 2)\sqrt{5} = 8\sqrt{5}$ is an example of the _____ property.

2. Like radicals have the same _____ of the same _____ .

3. $2\sqrt{5} + 7\sqrt{6}$ cannot be simplified because the _____ are different.

4. $3\sqrt[4]{6} - 5\sqrt[3]{6}$ cannot be simplified because the _____ are different.

Simplify and add or subtract wherever possible in exercises 5-24.

5. $7\sqrt{6} + 8\sqrt{6}$ 6. $10\sqrt{11} - 4\sqrt{11}$

7. $\sqrt{13} + 9\sqrt{13}$ 8. $17\sqrt{10} - \sqrt{10}$

9. $\sqrt{8} + \sqrt{8}$ 10. $\sqrt{45} + \sqrt{45}$

11. $9\sqrt{8} + 5\sqrt{2}$ 12. $2\sqrt{3} + 11\sqrt{27}$

13. $6\sqrt{50} - 2\sqrt{72}$ 14. $8\sqrt{18} - 4\sqrt{32}$

15. $-9\sqrt{54} + 2\sqrt{24}$

16. $2\sqrt{160} - 5\sqrt{40}$

17. $3\sqrt{5} - 3\sqrt{20} + 3\sqrt{45}$

18. $2\sqrt{11} + 2\sqrt{44} - 2\sqrt{99}$

19. $9\sqrt{32} - 6\sqrt{50} + 4\sqrt{162}$

20. $10\sqrt{175} - 2\sqrt{112} - 5\sqrt{63}$

21. $12\sqrt{48} - 5\sqrt{75} + 9\sqrt{243}$

22. $8\sqrt{96} + 11\sqrt{150} - 7\sqrt{54}$

23. $\dfrac{2}{9}\sqrt{486} - \dfrac{1}{4}\sqrt{96}$

24. $\dfrac{1}{5}\sqrt{175} + \dfrac{3}{4}\sqrt{112}$

25. Explain how the distributive property is actually used in this statement
$2\sqrt{3} + 4\sqrt{3} = 6\sqrt{3}$.

26. Explain why $\sqrt{11} + \sqrt{7}$ cannot be further simplified. Confirm, by using calculator approximations, that $\sqrt{11} + \sqrt{7}$ is *not* equal to $\sqrt{18}$.

Perform the indicated operations in exercises 27-44. Assume that all variables represent nonnegative real numbers.

27. $\sqrt{10} \cdot \sqrt{5} + 7\sqrt{2}$

28. $6\sqrt{14} \cdot \sqrt{2} + 9\sqrt{7}$

29. $\sqrt{16x} - \sqrt{81x} + \sqrt{4x}$

30. $\sqrt{25y} + \sqrt{100y} - \sqrt{49y}$

31. $\sqrt{48a^2} + a\sqrt{12}$

32. $\sqrt{45b^2} + b\sqrt{20}$

33. $3\sqrt{1000x^2} + 4x\sqrt{90} - 5x\sqrt{40}$

34. $5\sqrt{343m^2} - 8m\sqrt{112} + 7m\sqrt{175}$

35. $-12\sqrt{12k} + 9\sqrt{75k}$

36. $4\sqrt{27n} - 5\sqrt{108n}$

37. $6\sqrt{125ab^2} - 5b\sqrt{20a}$

38. $7\sqrt{98mn^2} - 9n\sqrt{32m}$

39. $5\sqrt[3]{16} - 6\sqrt[3]{54}$

40. $9\sqrt[3]{81} - 2\sqrt[3]{192}$

41. $9\sqrt[3]{64s^3r^2} - 4s\sqrt[3]{8r^2}$

42. $11\sqrt[3]{k^4} - 6k\sqrt[3]{27k}$

43. $2\sqrt[4]{81z^3} - 3\sqrt[4]{16z^3}$

44. $5n\sqrt[4]{625n} + 3\sqrt[4]{n^5}$

45. Describe in your own words how to add and subtract radicals.

46. In the directions for exercises 27–44, we made the assumption that all variables represent nonnegative real numbers. However, in Exercise 41 the variable actually *may* represent a negative number. Explain why this is so.

5.4 | **Simplifying Radical Expressions**

Here is a concise set of guidelines to follow when you are simplifying radical expressions. Although they are illustrated with square roots, the guidelines apply to higher roots as well.

Guideline for Simplifying Radical Expressions

1. If a radical represents a rational number, use that rational number in place of the radical.

   Examples: $\sqrt{36}$ is simplified by writing 6; $\sqrt{\dfrac{121}{9}}$ by writing $\dfrac{11}{3}$.

2. If a radical expression contains products of radicals, use the product rule for radicals, $\sqrt{x} \cdot \sqrt{y} = \sqrt{xy}$ to get a single radical.

   Examples: $\sqrt{3} \cdot \sqrt{5}$. is simplified to $\sqrt{15}$; $\sqrt{7} \cdot \sqrt{x}$ to $\sqrt{7x}$.

3. If a radicand has a factor that is a perfect square, express the radical as the product of the positive square root of the perfect square and the remaining radical factor.

   Examples: $\sqrt{12}$. is simplified to $\sqrt{12} = \sqrt{4 \cdot 3} = \sqrt{4} \cdot \sqrt{3} = 2\sqrt{3}$;

   $$\sqrt[3]{54} = \sqrt[3]{27 \cdot 2} = \sqrt[3]{27} \cdot \sqrt[3]{2} = 3\sqrt[3]{2} .$$

4. If a radical expression contains sums or differences of radicals, use the distributive property to combine like radicals.

   Examples: $5\sqrt{2} + 3\sqrt{2} = 8\sqrt{2}$, but $6\sqrt{3} + 2\sqrt{2}$ cannot be further simplified.

5. Rationalize any denominator containing a radical.

   Examples: $\dfrac{5}{\sqrt{7}}$ is rationalized as $\dfrac{5}{\sqrt{7}} = \dfrac{5 \cdot \sqrt{7}}{\sqrt{7} \cdot \sqrt{7}} = \dfrac{5\sqrt{7}}{7}$;

   $$\sqrt{\dfrac{5}{6}} = \dfrac{\sqrt{5}}{\sqrt{6}} = \dfrac{\sqrt{5} \cdot \sqrt{6}}{\sqrt{6} \cdot \sqrt{6}} = \dfrac{\sqrt{30}}{6} .$$

Simplify Products of Radical Expressions: Use these guidelines to find each product and simplify the answer.

EXAMPLE 1

(a) $\sqrt{3}\left(\sqrt{20} - \sqrt{45}\right) = \sqrt{3}\left(2\sqrt{5} - 3\sqrt{5}\right)$ Simplify inside the parentheses.

   $= \sqrt{3}\left(-\sqrt{5}\right)$ Subtract like radicals.

   $= -\sqrt{3 \cdot 5}$ Product rule; commutative property

   $= -\sqrt{15}$ Multiply.

(b) $\left(\sqrt{3} + 3\sqrt{7}\right)\left(\sqrt{3} - 5\sqrt{7}\right)$

The product of these sums of radicals can be found in the same way that we found the product of binomials in Chapter 3 using the FOIL method.

$$\begin{array}{cccc} \text{F} & \text{O} & \text{I} & \text{L} \end{array}$$
$$= \sqrt{3} \cdot \sqrt{3} + \sqrt{3}\left(-5\sqrt{7}\right) + 3\sqrt{7} \cdot \sqrt{3} + 3\sqrt{7}\left(-5\sqrt{7}\right)$$
$$= 3 - 5\sqrt{21} + 3\sqrt{21} - 15 \cdot 7 \qquad\qquad \text{Product rule}$$
$$= 3 - 2\sqrt{21} - 105 \qquad\qquad \text{Add like radicals.}$$
$$= -102 - 2\sqrt{21} \qquad\qquad \text{Combine terms.}$$

(c) $\sqrt[3]{2}\left(\sqrt[3]{12} - \sqrt[3]{20}\right) = \sqrt[3]{2} \cdot \sqrt[3]{12} - \sqrt[3]{2} \cdot \sqrt[3]{20} \qquad$ Distribute.
$$= \sqrt[3]{24} - \sqrt[3]{40} \qquad\qquad \text{Product rule}$$
$$= \sqrt[3]{8 \cdot 3} - \sqrt[3]{8 \cdot 5} \qquad\qquad \text{Factor.}$$
$$= \sqrt[3]{8} \cdot \sqrt[3]{3} - \sqrt[3]{8} \cdot \sqrt[3]{5} \qquad\qquad \text{Product rule}$$
$$= 2\sqrt[3]{3} - 2\sqrt[3]{5} \qquad\qquad \sqrt[3]{8} = 2$$

(d) $\left(\sqrt[3]{4} + \sqrt[3]{6}\right)\left(\sqrt[3]{9} + \sqrt[3]{12}\right) = \sqrt[3]{4} \cdot \sqrt[3]{9} + \sqrt[3]{4} \cdot \sqrt[3]{12} + \sqrt[3]{6} \cdot \sqrt[3]{9} + \sqrt[3]{6} \cdot \sqrt[3]{12}$
$$= \sqrt[3]{36} + \sqrt[3]{48} + \sqrt[3]{54} + \sqrt[3]{72}$$
$$= \sqrt[3]{36} + 2\sqrt[3]{6} + 3\sqrt[3]{2} + 2\sqrt[3]{9}$$

The rule for the special product, $(x - y)(x + y) = x^2 - y^2$, applies to radicals as well.

EXAMPLE 2

(a) $\left(5 - \sqrt{2}\right)\left(5 + \sqrt{2}\right) = 5^2 - \left(\sqrt{2}\right)^2 \qquad$ Special product. Let $a = 5$ and $b = \sqrt{2}$.
$$= 25 - 2 = 23 \qquad 5^2 = 25 \text{ and } \left(\sqrt{2}\right)^2 = 2$$

(b) $\left(\sqrt{11} + \sqrt{3}\right)\left(\sqrt{11} - \sqrt{3}\right) = \left(\sqrt{11}\right)^2 - \left(\sqrt{3}\right)^2 \qquad$ Special product. Let $a = \sqrt{11}$ and $b = \sqrt{3}$.
$$= 11 - 3 = 8 \qquad\qquad \left(\sqrt{11}\right)^2 = 11 \text{ and } \left(\sqrt{3}\right)^2 = 3$$

The pairs of expressions in Example 2, $5 - \sqrt{2}$ and $5 + \sqrt{2}$, and $\sqrt{11} + \sqrt{3}$ and $\sqrt{11} - \sqrt{3}$ are called **conjugates** of each other.

Self-Check 1
Find each product and simplify the answers.

1. $\sqrt{21}\left(\sqrt{3} + 2\sqrt{7}\right)$ 2. $\left(\sqrt{2} - 3\sqrt{7}\right)\left(2\sqrt{2} + 5\sqrt{7}\right)$

3. $\left(\sqrt{2} - 3\sqrt{7}\right)\left(2\sqrt{2} + 6\sqrt{7}\right)$ 4. $\left(2\sqrt{3} + 5\right)\left(2\sqrt{3} - 5\right)$

Simplify Quotients of Radical Expressions: We can use conjugates to simplify a radical expression that has terms in the denominator, where at least one of those terms is a square root radical. Multiply both the numerator and the denominator by the conjugate of the denominator.

EXAMPLE 3

(a) Simplify $\dfrac{6}{2+\sqrt{3}}$ by rationalizing the denominator.

We can eliminate the radical in the denominator by multiplying both numerator and denominator by $2-\sqrt{3}$, the conjugate of $2+\sqrt{3}$.

$$\frac{6}{2+\sqrt{3}} = \frac{6(2-\sqrt{3})}{(2+\sqrt{3})(2-\sqrt{3})} \qquad \text{Multiply by conjugate.}$$

$$= \frac{6(2-\sqrt{3})}{2^2-(\sqrt{3})^2} \qquad (x-y)(x+y) = x^2 - y^2$$

$$= \frac{6(2-\sqrt{3})}{4-3} \qquad 2^2 = 4 \text{ and } (\sqrt{3})^2 = 3$$

$$= \frac{6(2-\sqrt{3})}{1} \qquad \text{Subtract.}$$

$$= 6(2-\sqrt{3})$$

(b) Simplify $\dfrac{5+\sqrt{6}}{\sqrt{6}-4}$ by rationalizing the denominator.

Multiply numerator and denominator by $\sqrt{6}+4$.

$$\frac{5+\sqrt{6}}{\sqrt{6}-4} = \frac{(5+\sqrt{6})(\sqrt{6}+4)}{(\sqrt{6}-4)(\sqrt{6}+4)} \qquad \text{Multiply by conjugate.}$$

$$= \frac{5\sqrt{6}+20+6+4\sqrt{6}}{6-16} \qquad \text{FOIL}$$

$$= \frac{26+9\sqrt{6}}{-10} \qquad \text{Combine terms.}$$

$$= -\frac{26+9\sqrt{6}}{10} \qquad \frac{a}{-b} = -\frac{a}{b}$$

(c) Write $\dfrac{4\sqrt{5}+20}{16}$ in lowest terms.

Factor the numerator and denominator, and then divide numerator and denominator by any common factors.

$$\frac{4\sqrt{5}+20}{16} = \frac{4(\sqrt{5}+5)}{4 \cdot 4} = \frac{\sqrt{5}+5}{4}$$

Self-Check 2

Simplify. Rationalize the denominator, if necessary.

1. $\dfrac{10}{3-\sqrt{6}}$ 2. $\dfrac{2+\sqrt{5}}{\sqrt{5}+1}$ 3. $\dfrac{\sqrt{6}}{\sqrt{8}-\sqrt{6}}$ 4. $\dfrac{2\sqrt{27}-18}{12}$

Self-Check Answers

1.1 $3\sqrt{7} + 14\sqrt{3}$ **1.2** $-101 - \sqrt{14}$ **1.3** -122 **1.4** -13

2.1 $\dfrac{10\left(3 + \sqrt{6}\right)}{3}$ **2.2** $\dfrac{3 + \sqrt{5}}{4}$ **2.3** $3 + 2\sqrt{3}$ **2.4** $\dfrac{\sqrt{3} - 3}{2}$

5.4 EXERCISES

Perform the operations mentally in exercises 1-8, and write the answer without doing the intermediate steps.

1. $\sqrt{64} + \sqrt{9}$

2. $\sqrt{100} - \sqrt{49}$

3. $\sqrt{3} \cdot \sqrt{27}$

4. $\sqrt{11} \cdot \sqrt{11}$

5. $\sqrt{8}\left(\sqrt{32} - \sqrt{2}\right)$

6. $\sqrt{3}\left(\sqrt{27} + \sqrt{3}\right)$

7. $\sqrt[3]{27} - \sqrt[3]{8}$

8. $\sqrt[4]{64} - \sqrt[4]{16} - \sqrt{4}$

Simplify each expression in exercises 9-34. Use the guidelines.

9. $6\sqrt{3} + 5\sqrt{12}$

10. $9\sqrt{5} + 2\sqrt{80}$

11. $8\sqrt{96} - 10\sqrt{150}$

12. $2\sqrt{3200} - 3\sqrt{5000}$

13. $\sqrt{3}\left(\sqrt{2} - \sqrt{7}\right)$

14. $\sqrt{13}\left(\sqrt{11} + \sqrt{5}\right)$

15. $3\sqrt{5}\left(\sqrt{3} - 2\sqrt{5}\right)$

16. $6\sqrt{7}\left(\sqrt{7} + 5\sqrt{11}\right)$

17. $2\sqrt{10} \cdot \sqrt{2} - 7\sqrt{20}$

18. $4\sqrt{14} \cdot \sqrt{3} + 9\sqrt{42}$

19. $\left(2\sqrt{10} - 1\right)\left(3\sqrt{10} + 6\right)$

20. $\left(5\sqrt{14} - 7\right)\left(2\sqrt{14} + 9\right)$

21. $\left(3\sqrt{7} + 5\sqrt{2}\right)\left(2\sqrt{7} + \sqrt{2}\right)$

22. $\left(6\sqrt{10} + 2\sqrt{3}\right)\left(5\sqrt{10} - 7\sqrt{3}\right)$

23. $\left(3\sqrt{5} + 6\right)^2$

24. $\left(2\sqrt{6} - 5\right)^2$

25. $\left(2 - \sqrt{3}\right)\left(2 + \sqrt{3}\right)$

26. $\left(5 + \sqrt{7}\right)\left(5 - \sqrt{7}\right)$

27. $\left(\sqrt{10} - \sqrt{6}\right)\left(\sqrt{10} + \sqrt{6}\right)$

28. $\left(\sqrt{13} + \sqrt{7}\right)\left(\sqrt{13} - \sqrt{7}\right)$

29. $\left(\sqrt{2} - \sqrt{3}\right)\left(\sqrt{6} + \sqrt{2}\right)$

30. $\left(\sqrt{3} - \sqrt{5}\right)\left(\sqrt{15} + \sqrt{5}\right)$

31. $\left(\sqrt{12} - \sqrt{6}\right)\left(\sqrt{6} + \sqrt{24}\right)$

32. $\left(\sqrt{14} - \sqrt{7}\right)\left(\sqrt{7} + \sqrt{28}\right)$

33. $\left(6\sqrt{7} + 5\sqrt{5}\right)\left(2\sqrt{7} + 3\sqrt{5}\right)$

34. $\left(2\sqrt{2} - 4\sqrt{11}\right)\left(2\sqrt{2} - 7\sqrt{11}\right)$

35. In Example 1(b), the original expression simplifies to $-102 - 2\sqrt{21}$. Students often try to simplify expressions like this by combining the -102 and the -2 to get $-102 - 2\sqrt{21} = -104\sqrt{21}$, which is incorrect. Explain why.

36. If you try to rationalize the denominator of $\dfrac{3}{2 + \sqrt{3}}$ by multiplying the numerator and denominator by $2 + \sqrt{3}$, what problem arises? What should you multiply by?

Rationalize each denominator in exercises 37-48.

37. $\dfrac{2}{2 + \sqrt{3}}$

38. $\dfrac{62}{6 - \sqrt{5}}$

39. $\dfrac{12}{8 - \sqrt{6}}$

40. $\dfrac{10}{7 + \sqrt{5}}$

41. $\dfrac{\sqrt{3}}{2 + \sqrt{3}}$

42. $\dfrac{\sqrt{5}}{4 - \sqrt{5}}$

43. $\dfrac{\sqrt{3}}{\sqrt{5} + \sqrt{3}}$

44. $\dfrac{\sqrt{5}}{\sqrt{5} - \sqrt{3}}$

45. $\dfrac{\sqrt{20}}{\sqrt{5} + 2}$

46. $\dfrac{\sqrt{28}}{\sqrt{7} - 3}$

47. $\dfrac{\sqrt{2} + 3}{4 - \sqrt{5}}$

48. $\dfrac{2 - \sqrt{7}}{8 - \sqrt{2}}$

Write each quotient in lowest terms in exercises 49-54.

49. $\dfrac{8\sqrt{2} + 12}{4}$

50. $\dfrac{14 - 28\sqrt{2}}{7}$

51. $\dfrac{4\sqrt{7} - 8}{6}$

52. $\dfrac{5 - 15\sqrt{6}}{10}$

53. $\dfrac{15 - \sqrt{75}}{5}$

54. $\dfrac{3\sqrt{28} - 10}{10}$

In exercises 55-66, simplify each radical expression. Assume all variables represent nonnegative real numbers.

55. $\left(\sqrt{3x} + \sqrt{12}\right)\left(\sqrt{2x} + \sqrt{3}\right)$

56. $\left(\sqrt{6a} + \sqrt{20}\right)\left(\sqrt{2a} + \sqrt{5}\right)$

57. $\left(\sqrt{7y} + \sqrt{14}\right)\left(\sqrt{2y} - \sqrt{7}\right)$

58. $\left(2\sqrt{z} - \sqrt{2}\right)\left(3\sqrt{z} - \sqrt{5}\right)$

59. $\sqrt[3]{9}\left(\sqrt[3]{3} + 4\right)$

60. $\sqrt[3]{7}\left(\sqrt[3]{49} - 2\sqrt[3]{7}\right)$

61. $3\sqrt[4]{2}\left(3\sqrt[4]{4} - 3\sqrt[4]{8}\right)$

62. $5\sqrt[4]{3}\left(2\sqrt[4]{27} + 5\sqrt[4]{9}\right)$

63. $\left(\sqrt[3]{2} + \sqrt[3]{3}\right)\left(\sqrt[3]{2} - \sqrt[3]{3}\right)$

64. $\left(\sqrt[3]{4} - 2\sqrt[3]{5}\right)\left(\sqrt[3]{4} + 2\sqrt[3]{5}\right)$

65. $\left(\sqrt[3]{2} + \sqrt[3]{3}\right)\left(\sqrt[3]{4} - \sqrt[3]{6} + \sqrt[3]{9}\right)$

66. $\left(\sqrt[3]{4} - 2\sqrt[3]{5}\right)\left(\sqrt[3]{16} + 2\sqrt[3]{20} + 4\sqrt[3]{25}\right)$

5.5 | Fractional Exponents

Define a Fractional Exponent such as $a^{1/n}$: How should $3^{1/2}$ be defined? We want to define $3^{1/2}$ so that all the rules for exponents developed earlier in this book still hold. Then we should define $3^{1/2}$ so that

$$3^{1/2} \cdot 3^{1/2} = 3^{1/2 + 1/2} = 3^1 = 3.$$

This agrees with the product rule for exponents. By definition,

$$\left(\sqrt{3}\right)\left(\sqrt{3}\right) = 3.$$

Since both $3^{1/2} \cdot 3^{1/2}$ and $\sqrt{3} \cdot \sqrt{3}$ equal 3,

$$3^{1/2} \text{ should equal } \sqrt{3}.$$

Similarly,

$$3^{1/3} \text{ should equal } \sqrt[3]{3}.$$

These examples suggest the following definition.

Definition: $a^{1/n}$

If a is a nonnegative number and n is a positive integer,

$$a^{1/n} = \sqrt[n]{a}.$$

EXAMPLE 1
Simplify each expression by first writing it in radical form.

(a) $25^{1/2} = \sqrt{25} = 5$

(b) $8^{1/3} = \sqrt[3]{8} = 2$

(c) $(-27)^{1/3} = \sqrt[3]{-27} = -3$

(d) $-32^{1/5} = -\sqrt[5]{32} = -2$

Define $a^{m/n}$: Now a more general exponential expression like $8^{2/3}$ can be defined. By the power rule, $\left(a^m\right)^n = a^{mn}$, so that

$$8^{2/3} = \left(8^{1/3}\right)^2 = \left(\sqrt[3]{8}\right)^2 = 2^2 = 4.$$

However, $8^{2/3}$ could also be written as

$$8^{2/3} = \left(8^2\right)^{1/3} = 64^{1/3} = \sqrt[3]{64} = 4.$$

The expression can be evaluated either way to get the same answer. As the example suggests, taking the root first involves smaller numbers and is often easier. This example suggests the following definition for $a^{m/n}$.

Definition: $a^{m/n}$

If a is a nonnegative number and m and n is a positive integers, with $n > 0$,

$$a^{m/n} = \left(a^{1/n}\right)^m = \left(\sqrt[n]{a}\right)^m.$$

EXAMPLE 2
Evaluate each expression.

(a) $16^{3/2} = \left(16^{1/2}\right)^3 = 4^3 = 64$

(b) $32^{2/5} = \left(32^{1/5}\right)^2 = 2^2 = 4$

(c) $-81^{3/4} = -\left(81^{1/4}\right)^3 = -3^3 = -27$

Earlier, a^{-n} was defined as

$$a^{-n} = \frac{1}{a^n}$$

for nonzero numbers a and integers n. This same result applies for negative fractional exponents.

Definition: $a^{-m/n}$

If a is a nonnegative number and m and n is a positive integers, with $n > 0$,

$$a^{-m/n} = \frac{1}{a^{m/n}}.$$

EXAMPLE 3
Write each expression with a positive exponent and then evaluate.

(a) $64^{-2/3} = \dfrac{1}{64^{2/3}} = \dfrac{1}{\left(64^{1/3}\right)^2} = \dfrac{1}{4^2} = \dfrac{1}{16}$

(b) $3125^{-2/5} = \dfrac{1}{3125^{2/5}} = \dfrac{1}{\left(3125^{1/5}\right)^2} = \dfrac{1}{5^2} = \dfrac{1}{25}$

In Example 3(b), a common mistake is to write $3125^{-2/5} = -3125^{5/2}$. **This is incorrect.** The negative exponent does not indicate a negative number. Also, the negative exponent indicates the reciprocal of the *base,* not the reciprocal of the *exponent.*

Self-Check 1

Evaluate each expression.

1. $100^{1/2}$ 2. $-625^{3/4}$

3. $27^{-2/3}$ 4. $-125^{-2/3}$

Use Rules for Exponents with Fractional Exponents: All the rules for exponents given earlier still hold when the exponents are fractions. The next examples use these rules to simplify expressions with fractional exponents.

E X A M P L E 4

Simplify each expression. Write each answer in exponential form with only positive exponents.

(a) $6^{1/5} \cdot 6^{2/5} = 6^{1/5+2/5} = 6^{3/5}$

(b) $\dfrac{2^{1/6}}{2^{5/6}} = 2^{1/6-5/6} = 2^{-4/6} = 2^{-2/3} = \dfrac{1}{2^{2/3}}$

(c) $\left(4^{1/4}\right)^2 = 4^{2(1/4)} = 4^{1/2} = \sqrt{4} = 2$

(d) $\dfrac{3^{-1/3} \cdot 3^{-1}}{3^{2/3}} = \dfrac{3^{-1/3+(-1)}}{3^{2/3}} = \dfrac{3^{-4/3}}{3^{2/3}} = 3^{-4/3-2/3} = 3^{-6/3} = 3^{-2} = \dfrac{1}{3^2} = \dfrac{1}{9}$

(e) $x^{4/7} \cdot x^{1/7} = x^{4/7+1/7} = x^{5/7}$

(f) $\dfrac{y^{8/9}}{y^{1/3}} = y^{8/9-1/3} = y^{8/9-3/9} = y^{5/9}$

(g) $\left(x^3 y^{2/3}\right)^6 = \left(x^3\right)^6 \left(y^{2/3}\right)^6 = x^{18} y^4$

(h) $\left(\dfrac{a^{2/3}}{b^{5/8}}\right)^{12} = \dfrac{\left(a^{2/3}\right)^{12}}{\left(b^{5/8}\right)^{12}} = \dfrac{a^8}{b^{15/2}}, \ b > 0$

Use Fractional Exponents to Simplify Radicals: Sometimes it is easier to simplify a radical by first writing it in exponential form.

E X A M P L E 5

(a) $\sqrt[8]{16^4} = \left(16^4\right)^{1/8} = 16^{4/8} = 16^{1/2} = \sqrt{16} = 4$

(b) $\left(\sqrt[12]{x}\right)^6 = \left(x^{1/12}\right)^6 = x^{6/12} = x^{1/2} = \sqrt{x}$

   Here it is assumes that $x \geq 0$.

Self-Check 2

Simplify each expression. Write each answer in exponential form with only positive exponents.

1. $9^{1/6} \cdot 9^{1/3}$ 2. $\left(3^{3/4}\right)^2$

3. $\left(a^2 b^{3/5}\right)^{10}$ 4. $\sqrt[12]{64^4}$

Self-Check Answers

1.1 10	**1.2** -125	**1.3** $\dfrac{1}{9}$	**1.4** $-\dfrac{1}{25}$
2.1 3	**2.2** $3\sqrt{3}$	**2.3** $a^{20}b^6$	**2.4** 4

5.5 EXERCISES

For exercises 1-4, decide which one of the four choices is not *equal to the given expression.*

1. $81^{1/2}$ (a) -9 (b) 9 (c) $\sqrt{81}$ (d) $81^{0.5}$

2. $16^{1/4}$ (a) 2 (b) $\sqrt[4]{16}$ (c) $16^{0.25}$ (d) $\dfrac{16}{4}$

3. $-27^{1/3}$ (a) $-\sqrt{9}$ (b) -3 (c) 3 (d) $-\sqrt[3]{27}$

4. $-81^{1/4}$ (a) $-\sqrt{9}$ (b) -3 (c) 3 (d) $-\sqrt[4]{81}$

Simplify each expression by first writing it in radical form in exercises 5-28.

5. $36^{1/2}$ 6. $144^{1/2}$ 7. $125^{1/3}$ 8. $1000^{1/3}$

9. $81^{1/4}$ 10. $16^{1/4}$ 11. $243^{1/5}$ 12. $1024^{1/5}$

13. $4^{5/2}$ 14. $16^{3/2}$ 15. $8^{5/3}$ 16. $64^{2/3}$

17. $32^{3/5}$ 18. $64^{5/6}$ 19. $-81^{1/2}$ 20. $-144^{3/2}$

21. $-27^{4/3}$ 22. $-1000^{5/3}$ 23. $216^{-2/3}$ 24. $32^{-3/5}$

25. $81^{-1/4}$ 26. $128^{-2/7}$ 27. $-125^{-1/3}$ 28. $-16^{-3/4}$

Simplify each expression in exercises 29-52. Write answers in exponential form with only positive exponents. Assume that all variables represent positive numbers.

29. $3^{1/3}\cdot 3^{5/3}$ 30. $7^{3/4}\cdot 7^{5/4}$ 31. $9^{1/4}\cdot 9^{-3/4}$ 32. $8^{1/3}\cdot 8^{-2/3}$

33. $\dfrac{10^{3/4}}{10^{5/4}}$ 34. $\dfrac{6^{5/6}}{6^{1/6}}$ 35. $\dfrac{5^{-2/5}}{5^{-3/5}}$ 36. $\dfrac{7^{-3/7}}{7^{-4/7}}$

37. $\left(6^{2/7}\right)^{7}$ 38. $\left(3^{4/9}\right)^{9}$ 39. $\left(14^{5/6}\right)^{3/5}$ 40. $\left(15^{1/6}\right)^{2/7}$

41. $\left(\dfrac{36}{49}\right)^{3/2}$ 42. $\left(\dfrac{27}{1000}\right)^{2/3}$ 43. $\dfrac{3^{5/6}\cdot 3^{-2/5}}{3^{1/5}}$ 44. $\dfrac{2^{5/4}\cdot 2^{-1/4}}{2^{-3/4}}$

45. $\dfrac{7^{2/3}\cdot 7^{-8/3}}{7^{-5/3}}$ 46. $\dfrac{5^{-5/9}\cdot 5^{11/9}}{5^{8/9}}$ 47. $\dfrac{x^{4/5}}{x^{-4/5}}$ 48. $\dfrac{y^{-4/6}}{y^{-1/6}}$

49. $\left(a^{7}b^{2/5}\right)^{10/7}$ 50. $\left(s^{5}t^{5/4}\right)^{2/5}$ 51. $\left(\dfrac{x^{2/3}}{y^{4/5}}\right)^{6/7}$ 52. $\left(\dfrac{m^{3/5}}{n^{5/7}}\right)^{7/9}$

In exercises 53-60, simplify each radical by first writing it in exponential form. Give the answer as an integer or a radical in simplest form. Assume that all variables represent nonnegative numbers.

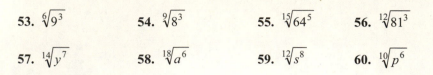

53. $\sqrt[6]{9^3}$ **54.** $\sqrt[9]{8^3}$ **55.** $\sqrt[15]{64^5}$ **56.** $\sqrt[12]{81^3}$

57. $\sqrt[14]{y^7}$ **58.** $\sqrt[18]{a^6}$ **59.** $\sqrt[12]{s^8}$ **60.** $\sqrt[10]{p^6}$

5.6 | Complex Numbers

As discussed in Chapter 1, the set of real numbers includes many other number sets (the rational numbers, integers, and natural numbers, for example). In this section a new set of numbers is introduced that includes the set of real numbers, as well as numbers that are even roots of negative numbers, like $\sqrt{-2}$.

Simplify Square Roots of Negative Numbers: The equation $x^2 + 1 = 0$ has no real number solutions, since any solution must be a number whose square is -1. To provide a solution for the equation $x^2 + 1 = 0$, a new number i is defined so that

$$i^2 = -1.$$

That is, i is a number whose square is -1, so $i = \sqrt{-1}$. This definition of i makes it possible to define any square root of a negative number as follows.

For any positive real number b, $\sqrt{-b} = i\sqrt{b}$.

EXAMPLE 1
Write each number as a product of a real number and i.

(a) $\sqrt{-49} = i\sqrt{49} = 7i$

(b) $\sqrt{-10} = i\sqrt{10} = i\sqrt{10}$

It is easy to mistake $\sqrt{10}i$ for $\sqrt{10i}$, with the i under the radical. For this reason, we often write $\sqrt{10}i$ as $i\sqrt{10}$.

When finding a product such as $\sqrt{-4} \cdot \sqrt{-25}$, we *cannot* use the product rule for radicals since that rule applies only when both radicals represent real numbers. For this reason, we *always* change $\sqrt{-b}$ $(b > 0)$ to the form $i\sqrt{b}$ before performing any multiplications or divisions. For example,

$$\sqrt{-4} \cdot \sqrt{-25} = i\sqrt{4} \cdot i\sqrt{25} = i \cdot 2 \cdot i \cdot 5 = 10i^2.$$

Since $i^2 = -1$,

$$10i^2 = 10(-1) = -10.$$

Using the product rule for radicals incorrectly gives a wrong answer.

$$\sqrt{-4} \cdot \sqrt{-25} = \sqrt{(-4)(-25)} = \sqrt{100} = 10 \quad \text{INCORRECT}$$

EXAMPLE 2
Multiply.

(a) $\sqrt{-5} \cdot \sqrt{-6} = i\sqrt{5} \cdot i\sqrt{6} = i^2\sqrt{5 \cdot 6} = (-1)\sqrt{30} = -\sqrt{30}$

(b) $\sqrt{-3} \cdot \sqrt{-27} = i\sqrt{3} \cdot i\sqrt{27} = i^2\sqrt{3 \cdot 27} = (-1)\sqrt{81} = -9$

(c) $\sqrt{-4} \cdot \sqrt{-25} = i\sqrt{4} \cdot i\sqrt{25} = i \cdot 2 \cdot i \cdot 5 = 10i^2 = -10$

The methods used to find products also apply to quotients.

EXAMPLE 3
Divide.

(a) $\dfrac{\sqrt{-8}}{\sqrt{-2}} = \dfrac{i\sqrt{8}}{i\sqrt{2}} = \sqrt{\dfrac{8}{2}} = \sqrt{4} = 2$

(b) $\dfrac{\sqrt{-54}}{\sqrt{6}} = \dfrac{i\sqrt{54}}{\sqrt{6}} = i\sqrt{\dfrac{54}{6}} = i\sqrt{9} = 3i$

Self-Check 1

Simplify

1. $\sqrt{-8}$ 2. $\sqrt{-6} \cdot \sqrt{-6}$ 3. $\sqrt{-12} \cdot \sqrt{-27}$ 4. $\dfrac{\sqrt{-480}}{\sqrt{-8}}$

Define Complex Numbers: With the new number i and the real numbers, a new set of numbers can be formed that includes the real numbers as a subset. The *complex numbers* are defined as follows.

Definition: Complex Number

If a and b are real numbers, then any number of the form $a + bi$ is called a **complex number**.

In the complex number $a + bi$, the number a is called the **real part** and b is called the **imaginary part**. When $b = 0$, $a + bi$ is a real number, so the real numbers are a subset of the complex numbers. Complex numbers with $b \neq 0$ are called **imaginary numbers**.* In spite of their name, imaginary numbers are very useful in applications, particularly in work with electricity.

The relationships among the various sets of numbers discussed in this book are shown in Figure 1.

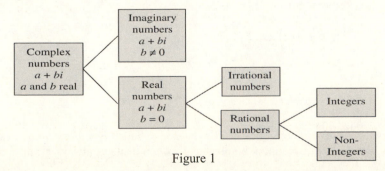

Figure 1

* Some texts define bi as the imaginary part of the complex number $a + bi$. Also, imaginary numbers are sometimes defined as complex numbers with $a = 0$ and $b \neq 0$.

Add and Subtract Complex Numbers: The commutative, associative, and distributive properties for real numbers are also valid for complex numbers. To add complex numbers, add their real parts and add their imaginary parts.

E X A M P L E 4

(a) $(3 + 4i) + (6 + 2i) = (3 + 6) + (4 + 2)i = 9 + 8i$

   Use the commutative and associative properties as well as the distributive property.

(b) $5 + (8 - 4i) = (5 + 8) - 4i = 13 - 4i$

To subtract complex numbers, subtract their real parts and subtract their imaginary parts.

E X A M P L E 5

(a) $(6 - 4i) - (8 + 2i) = (6 - 8) + (-4 - 2)i = -2 - 6i$

(b) $(4 - 6i) - (-1 - 8i) = [4 - (-1)] + [-6 - (-8)]i = 5 + 2i$

As in Example 5(b), a complex number written in the form $a + bi$, like $5 + 2i$, is in **standard form**.

Self-Check 2
Simplify
1. $(3 - 2i) + (6 + 5i)$ **2.** $(5 + 3i) + (10 - 3i)$
3. $(2 - i) - (5 + 8i)$ **4.** $(3 - 2i) - (3 + 5i)$

Multiply Complex Numbers: Complex numbers of the form $a + bi$ have the same form as a binomial, so we multiply two complex numbers by using the FOIL method for multiplying binomials. (Recall that FOIL stands for *First-Outside-Inside-Last*.)

E X A M P L E 6

Multiply $3 + 2i$ and $5 - 6i$.

$$(3 + 2i)(5 - 6i) = 3(5) + 3(-6i) + 2i(5) + 2i(-6i)$$

$$\quad\quad\quad\quad\quad\quad \uparrow \quad\quad \uparrow \quad\quad \uparrow \quad\quad \uparrow$$
$$\quad\quad\quad\quad\quad\quad \text{First} \;\; \text{Outside} \;\; \text{Inside} \;\; \text{Last}$$

Now simplify. (Remember that $i^2 = -1$.)

$$(3 + 2i)(5 - 6i) = 15 - 18i + 10i - 12i^2$$
$$= 15 - 8i - 12(-1) \quad\quad \text{Let } i^2 = -1.$$
$$= 15 - 8i + 12$$
$$= 27 - 8i$$

The two complex numbers $a + bi$ and $a - bi$ are called *complex conjugates* of each other. The product of a complex number and its conjugate is always a real number, as shown.

$$(a + bi)(a - bi) = a \cdot a + a(-bi) + bi(a) + bi(-bi)$$
$$= a^2 - abi + abi - b^2 i^2$$
$$= a^2 - b^2(-1)$$
$$(a + bi)(a - bi) = a^2 + b^2$$

For example, $(3 + 4i)(3 - 4i) = 3^2 + 4^2 = 9 + 16 = 25$.

Divide Complex Numbers: The quotient of two complex numbers should be a complex number. To write the quotient as a complex number, we need to eliminate i in the denominator. We use conjugates to do this.

EXAMPLE 7

Divide $10 + 15i$ and $4 - 7i$.

Multiply the numerator and the denominator of the quotient by the conjugate of the denominator. The conjugate of $4 - 7i$ is $4 + 7i$.

$$\frac{(10 + 15i)(4 + 7i)}{(4 - 7i)(4 + 7i)} = \frac{40 + 70i + 60i + 105i^2}{4^2 + 7^2}$$
$$= \frac{40 + 130i - 105}{16 + 49}$$
$$= \frac{-65 + 130i}{65}$$
$$= \frac{65(-1 + 2i)}{65}$$
$$= -1 + 2i$$

Notice that this is just like rationalizing the denominator. The final result is given in standard form.

Self-Check 3

Simplify.

1. $(3 - 5i)(5 - 7i)$

2. $\left(6 + 2i\sqrt{2}\right)\left(3 - i\sqrt{2}\right)$

3. $\dfrac{6 + 7i}{4 - 2i}$

4. $\dfrac{5 - 4i}{3 + 2i}$

Powers of i: Because i^2 is defined to be -1, we can find higher powers of i, as shown in the following examples.

Powers of i:

$$i^1 = i \qquad\qquad i^5 = i \cdot i^4 = i \cdot 1 = i$$
$$i^2 = -1 \qquad\qquad i^6 = i^2 \cdot i^4 = -1 \cdot 1 = -1$$
$$i^3 = i \cdot i^2 = i(-1) = -i \qquad\qquad i^7 = i^3 \cdot i^4 = -i \cdot 1 = -i$$
$$i^4 = i^2 \cdot i^2 = (-1)(-1) = 1 \qquad i^8 = i^4 \cdot i^4 = 1 \cdot 1 = 1$$

As these examples suggest, the powers of i rotate through the four numbers $i, -1, -i$, and 1. Larger powers of i can be simplified by using the fact that $i^4 = 1$. For example,

$$i^{79} = \left(i^4\right)^{19} \cdot i^3 = 1^{19} \cdot i^3 = i^3 = -i.$$

This example suggests a quick method for simplifying large powers of i.

EXAMPLE 8

(a) $i^{20} = \left(i^4\right)^5 = 1^5 = 1$

(b) $i^{41} = i^{40} \cdot i^1 = \left(i^4\right)^{10} \cdot i = 1^{10} \cdot i = 1 \cdot i = i$

(c) $i^{-2} = \dfrac{1}{i^2} = \dfrac{1}{-1} = -1$

Self-Check 4
Simplify.
1. i^{14} **2.** i^{101} **3.** i^{403} **4.** i^{-8}

Self-Check Answers

1.1 $2i\sqrt{2}$	1.2 -6	1.3 -18	1.4 $2\sqrt{15}$
2.1 $9 + 3i$	2.2 15 or $15 + 0i$	2.3 $-3 - 9i$	2.4 $-7i$ or $0 - 7i$
3.1 $-20 - 46i$	3.2 22 or $22 + 0i$	3.3 $\dfrac{1}{2} + 2i$	3.4 $\dfrac{7}{13} - \dfrac{22}{13}i$
4.1 -1	4.2 i	4.3 $-i$	4.4 1

5.6 EXERCISES

In exercises 1-6, decide whether each expression is equal to 1, -1, i, or $-i$.

1. $\sqrt{-1}$ **2.** $-\sqrt{-1}$ **3.** $-i^2$

4. i^{-2} **5.** $(-i)^3$ **6.** i^{-4}

In exercises 7-14, write each number as a product of a real number and i. Simplify all radical expressions.

7. $\sqrt{-144}$ **8.** $\sqrt{-81}$ **9.** $-\sqrt{-169}$ **10.** $-\sqrt{-225}$

11. $\sqrt{-10}$ **12.** $\sqrt{-35}$ **13.** $\sqrt{-20}$ **14.** $\sqrt{-75}$

Multiply or divide as indicated in exercises 15-22.

15. $\sqrt{-14} \cdot \sqrt{-14}$ **16.** $\sqrt{-21} \cdot \sqrt{-21}$ **17.** $\sqrt{-9} \cdot \sqrt{-49}$ **18.** $\sqrt{-100} \cdot \sqrt{-81}$

19. $\dfrac{\sqrt{-120}}{\sqrt{-40}}$ **20.** $\dfrac{\sqrt{-130}}{\sqrt{-65}}$ **21.** $\dfrac{\sqrt{-12}}{\sqrt{3}}$ **22.** $\dfrac{\sqrt{-20}}{\sqrt{5}}$

23. (a) Every real number is a complex number. Explain why this is so.
   (b) Not every complex number is a real number. Give an example of this, and explain why this statement is true.

24. Explain how to perform addition, subtraction, multiplication, and division with complex numbers. Give examples.

Add or subtract as indicated in exercises 25-38. Write your answers in the form a + bi.

25. $(2 + 9i) + (3 - 6i)$ **26.** $(3 - 7i) + (5 - 2i)$

27. $(-11 - i) + (1 - 2i)$ **28.** $(7 - 5i) + (-12 - 16i)$

29. $(8 + 7i) - (5 - 2i)$ **30.** $(13 - 9i) - (4 + 3i)$

31. $(16 - 14i) - (-7 - 12i)$ **32.** $(9 + i) - (6 - i)$

33. $(7 - 2i) - (-5 - 2i)$ **34.** $(-10 + 3i) - (7 + 3i)$

35. $(12 - 8i) + (5 - 4i) + (8 + 16i)$ **36.** $(13 - 6i) + (-8 - 14i) + (12 - i)$

37. $\left[(7 - 8i) - (12 + i)\right] + (3 + 6i)$ **38.** $\left[(-11 + 8i) - (-4 + 2i)\right] - (7 + 2i)$

39. Fill in the blank with the correct response:
   Because $(7 + 6i) - (2 - i) = 5 + 7i$, using the definition of subtraction, we can check this to find that $(5 + 7i) + (2 - i) =$ _____.

40. Fill in the blank with the correct response:
   Because $\dfrac{-26}{2 - 3i} = -4 - 6i$, using the definition of division, we can check this to find that $(-4 - 6i) \cdot (2 - 3i) =$ _____.

Multiply in exercises 41-56.

41. $(2i)(7i)$ **42.** $(9i)(10i)$

43. $(-4i)(-3i)$ **44.** $(-12i)(-5i)$

45. $2i(-2 + 9i)$ **46.** $3i(4 - 5i)$

47. $(2 + 9i)(4 - 5i)$ **48.** $(6 - i)(5 + 7i)$

49. $(2 - 7i)^2$ **50.** $(6 + 5i)^2$

51. $2i(-1 + i)^2$ **52.** $3i(2 - i)^2$

53. $(7 + 6i)(7 - 6i)$ **54.** $(5 - 11i)(5 + 11i)$

55. $(6 - i\sqrt{3})(6 + i\sqrt{3})$ **56.** $(7 + i\sqrt{5})(7 - i\sqrt{5})$

Write each expression in the form a + bi in exercises 57-64.

57. $\dfrac{34}{3 + 5i}$ **58.** $\dfrac{130}{8 - i}$ **59.** $\dfrac{6 + 2i}{2 - i}$ **60.** $\dfrac{-50 - 20i}{3 - i}$

61. $\dfrac{68i}{3 - 5i}$ **62.** $\dfrac{-106i}{2 - 7i}$ **63.** $\dfrac{6 - 4i}{5 - 2i}$ **64.** $\dfrac{-1 + 9i}{6 + 4i}$

Find each power of i in exercises 65-70.

65. i^{22} **66.** i^{38} **67.** i^{99}

68. i^{121} **69.** i^{-5} **70.** i^{-7}

71. A student simplified i^{-22} as follows:

$$i^{-22} = i^{-22} \cdot i^{24} = i^{-22+24} = i^{2} = -1.$$

Explain the mathematical justification for this correct work.

72. Explain why $(2 + 9i)(7 - 16i)$ and $(2 + 9i)(7 - 16i)(i^{120})$ must be equal. (Do not actually perform the computations.)

CH 5 | Summary

KEY TERMS

5.1 square roots   perfect square
   radicand   cube root
   radical   fourth root
   radical sign   index (order)
   radical expression   principal root

5.2 product rule for radicals   rationalizing the denominator
   quotient rule for radicals   simplified form
   perfect cube

5.3 like radicals

5.4 conjugate

5.5 fractional exponent

5.6 complex number   imaginary numbers
   real part   standard form (of a complex number)
   imaginary part

CH 5 | Quick Review

5.1 EVALUATING ROOTS

If a is a positive real number, $\sqrt{a}$ is the positive square root of a ;

$-\sqrt{a}$ is the negative square root of a ;

$\sqrt{0} = 0$.

If a is a negative real number, $\sqrt{a}$ is not a real number.

If a is a positive rational number, $\sqrt{a}$ is rational if a is a perfect square.

$\sqrt{a}$ is irrational if a is not a perfect square.

Each real number has exactly one real cube root.

5.2 MULTIPLICATION AND DIVISION OF RADICALS

Product Rule for Radicals

For nonnegative real numbers x and y ,

$$\sqrt{x} \cdot \sqrt{y} = \sqrt{xy} \text{ and } \sqrt{xy} = \sqrt{x} \cdot \sqrt{y}.$$

Quotient Rule for Radicals

If x and y are nonnegative real numbers and y is not 0,

$$\frac{\sqrt{x}}{\sqrt{y}} = \sqrt{\frac{x}{y}} \text{ and } \sqrt{\frac{x}{y}} = \frac{\sqrt{x}}{\sqrt{y}}.$$

If all indicated roots are real,

$$\sqrt[n]{x} \cdot \sqrt[n]{y} = \sqrt[n]{xy} \text{ and } \frac{\sqrt[n]{x}}{\sqrt[n]{y}} = \sqrt[n]{\frac{x}{y}}, y \neq 0.$$

The denominator of a radical can be rationalized by multiplying both the numerator and denominator by the same number.

5.3 ADDITION AND SUBTRACTION OF RADICALS

Add and subtract like radicals by using the distributive property. Only like radicals can be combined in this way.

5.4 SIMPLIFYING RADICAL EXPRESSIONS

When appropriate, use the rules for adding and multiplying polynomials to simplify radical expressions.

Any denominator with radicals should be rationalized.

If a radical expression contains two terms in the denominator and at least one of those terms is a radical, multiply both the numerator and denominator by the conjugate of the denominator.

5.5 FRACTIONAL EXPONENTS

Assume $a \geq 0$, m and n are integers, $n \geq 0$.

$$a^{1/n} = \sqrt[n]{a} \text{ and } a^{m/n} = \sqrt[n]{a^m} = \left(\sqrt[n]{a}\right)^m \text{ and } a^{-m/n} = \frac{1}{a^{m/n}} \ (a \neq 0)$$

5.6 COMPLEX NUMBERS

$$i^2 = -1$$

For any positive number b,

$$\sqrt{-b} = i\sqrt{b}.$$

To multiply $\sqrt{-2} \cdot \sqrt{-18}$, first change each factor to the form $i\sqrt{b}$, then multiply. The same procedure applies to a quotient such as $\dfrac{\sqrt{-18}}{\sqrt{-2}}$.

Adding and Subtracting Complex Numbers

$$(a + bi) + (c + di) = (a + c) + (b + d)i$$
$$(a + bi) - (c + di) = (a - c) + (b - d)i$$

Multiplying and Dividing Complex Numbers
Multiply complex numbers by using the FOIL method.

Divide complex numbers by multiplying the numerator and the denominator by the complex conjugate of the denominator.

CH 5 Review Exercises

In exercises 1-6, find all square roots of each number.

1. 1   **2.** 4   **3.** 64

4. 81   **5.** 576   **6.** 441

Find each indicated root in exercises 7-14. If the root is not a real number, say so.

7. $\sqrt{25}$ **8.** $-\sqrt{121}$ **9.** $\sqrt[3]{125}$ **10.** $\sqrt[4]{10,000}$

11. $\sqrt{-1600}$ **12.** $-\sqrt{4900}$ **13.** $\sqrt{\dfrac{81}{289}}$ **14.** $\sqrt{\dfrac{324}{361}}$

15. If $\sqrt[4]{x}$ is a real number, then what kind of number must x be?

16. Find the value of x in the figure.

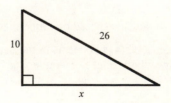

Determine whether each number is rational, irrational, *or* not a real number. *If the number is rational, give its exact value. If the number is irrational, give its decimal approximation rounded to the nearest thousandth.*

17. $\sqrt{26}$ **18.** $\sqrt{841}$ **19.** $-\sqrt{784}$ **20.** $\sqrt{-9}$

Use the product rule to simplify each expression in exercises 21-28.

21. $\sqrt{10} \cdot \sqrt{14}$ **22.** $-\sqrt{72}$ **23.** $\sqrt{250}$ **24.** $\sqrt[3]{-64}$

25. $\sqrt[3]{3375}$ **26.** $\sqrt{24} \cdot \sqrt{54}$ **27.** $\sqrt{135} \cdot \sqrt{10}$ **28.** $\sqrt{15} \cdot \sqrt{27}$

Use the product rule, the quotient rule, or both to simplify each expression in exercises 29-36.

29. $-\sqrt{\dfrac{169}{361}}$ **30.** $\sqrt{\dfrac{5}{64}}$ **31.** $\sqrt{\dfrac{7}{121}}$ **32.** $\sqrt{\dfrac{3}{5}} \cdot \sqrt{\dfrac{7}{5}}$

33. $\sqrt{\dfrac{7}{2}} \cdot \sqrt{\dfrac{5}{8}}$ **34.** $\dfrac{5\sqrt{21}}{\sqrt{7}}$ **35.** $\dfrac{38\sqrt{20}}{19\sqrt{5}}$ **36.** $\dfrac{10\sqrt{120}}{2\sqrt{60}}$

Simplify each expression in exercises 37-42. Assume that all variables represent nonnegative real numbers.

37. $\sqrt{d} \cdot \sqrt{d}$ **38.** $\sqrt{x} \cdot \sqrt{y}$ **39.** $\sqrt{a^{12}}$

40. $\sqrt{s^{10}t^{16}}$ **41.** $\sqrt{a^{13}b^{17}}$ **42.** $\sqrt{169x^{12}y^{13}}$

43. Use a calculator to find approximations for $\sqrt{0.2}$ and $\dfrac{\sqrt{5}}{5}$. Based on your results, do you think that these two expressions represent the same number? If so, verify it algebraically.

Simplify and combine terms where possible in exercises 44-51.

44. $3\sqrt{7} + 16\sqrt{7}$ **45.** $8\sqrt{27} - 2\sqrt{27}$

46. $16\sqrt{12} + 9\sqrt{27}$ **47.** $10\sqrt{27} - 5\sqrt{75} + 2\sqrt{147}$

48. $12\sqrt{5} - 9\sqrt{20} + 5\sqrt{125}$ **49.** $\dfrac{4}{5}\sqrt{50} - \dfrac{2}{3}\sqrt{18}$

50. $\dfrac{10}{3}\sqrt{54} + \dfrac{7}{2}\sqrt{24}$ **51.** $\sqrt{15} \cdot \sqrt{6} - 7\sqrt{40} + \sqrt{1000}$

Simplify each expression in exercises 52-55. Assume that all variables represent nonnegative real numbers.

52. $\sqrt{36x} + \sqrt{49x} - \sqrt{100x}$ **53.** $\sqrt{25y} - 6\sqrt{y} + 2\sqrt{121y}$

54. $\sqrt{72s^2} - s\sqrt{50}$ **55.** $3g\sqrt{98g^2h} - 5g^2\sqrt{8h}$

Perform the indicated operations and write answers in simplest form in exercises 56-65. Assume that all variables represent positive real numbers.

56. $\dfrac{5\sqrt{10}}{\sqrt{7}}$ **57.** $\dfrac{3}{\sqrt{3}}$ **58.** $\dfrac{20}{\sqrt{48}}$ **59.** $\dfrac{\sqrt{3}}{\sqrt{13}}$ **60.** $\sqrt{\dfrac{3}{5}}$

61. $\sqrt{\dfrac{6}{13}} \cdot \sqrt{26}$ **62.** $\sqrt{\dfrac{1}{5}} \cdot \sqrt{\dfrac{2}{7}}$ **63.** $\sqrt{\dfrac{x^2}{25y}}$ **64.** $\sqrt{\dfrac{a^3}{6b}}$ **65.** $\sqrt{\dfrac{s}{12t^3}}$

66. Explain how you would show, without using a calculator, that $\dfrac{\sqrt{10}}{4}$ and $\sqrt{\dfrac{80}{128}}$ represent the exact same number. Perform the actual necessary steps.

Simplify each expression in exercises 67-72.

67. $-\sqrt{5}\left(\sqrt{3} + \sqrt{20}\right)$

68. $3\sqrt{5}\left(2\sqrt{5} - \sqrt{10}\right)$

69. $\left(2\sqrt{7} - 5\right)\left(3\sqrt{7} + 2\right)$

70. $\left(5\sqrt{3} + 2\right)^2$

71. $\left(3\sqrt{5} + 7\sqrt{2}\right)\left(3\sqrt{5} - 7\sqrt{2}\right)$

72. $\left(5\sqrt{11} - 2\sqrt{3}\right)\left(5\sqrt{11} + 2\sqrt{3}\right)$

Rationalize each denominator in exercises 73-78.

73. $\dfrac{3}{5 + \sqrt{3}}$

74. $\dfrac{5}{1 - \sqrt{7}}$

75. $\dfrac{\sqrt{10}}{\sqrt{5} + 3}$

76. $\dfrac{\sqrt{5}}{1 - \sqrt{5}}$

77. $\dfrac{\sqrt{3} + 1}{\sqrt{5} - 3}$

78. $\dfrac{1 - \sqrt{2}}{6 - \sqrt{3}}$

Write each quotient in lowest terms in exercises 79-81.

79. $\dfrac{16 + 10\sqrt{2}}{4}$

80. $\dfrac{12 - \sqrt{432}}{24}$

81. $\dfrac{\sqrt{125} - \sqrt{50}}{25}$

Simplify each expression in exercises 82-87. Assume that all variables represent positive real numbers.

82. $49^{\frac{1}{2}}$

83. $-8^{\frac{2}{3}}$

84. $5^{\frac{2}{3}} \cdot 5^{\frac{4}{3}}$

85. $\dfrac{11^{\frac{5}{7}}}{11^{\frac{12}{7}}}$

86. $\dfrac{x^{\frac{6}{5}} \cdot x^{-\frac{7}{5}}}{x^{-\frac{4}{5}}}$

87. $\sqrt[24]{169^{12}}$

88. Answer true or false to each of the following

    (a) $i^3 = -i$ (b) $i = -1$ (c) $i = \sqrt{-1}$

    (d) $i^{41} = i$ (e) $\sqrt{-4} = 2i$ (f) $\sqrt{-5} \cdot \sqrt{-6} = \sqrt{30}$

Perform the indicated operations in exercises 89-94. Express the answers in the form a + bi.

89. $(2 + 8i) + (7 - 9i)$

90. $(-8 - 6i) - (2 - 11i)$

91. $5i(2 + 5i)$

92. $(3 + 4i)(5 - 6i)$

93. $(9 - i)(9 + i)$

94. $\dfrac{6 + i}{1 - 4i}$

CH 5 | **Test**

On this test, assume that all variables represent positive real numbers.

1. Find all square roots of 324.

Simplify where possible.

2. $\sqrt[3]{729}$

3. $-\sqrt{7500}$

4. $\sqrt{\dfrac{50}{27}}$

5. $\sqrt[4]{32}$

6. $\dfrac{15\sqrt{21}}{3\sqrt{7}}$

7. $2\sqrt{20} - 8\sqrt{80}$

8. $20x\sqrt{27} - 90\sqrt{12x^2} + 3x\sqrt{300}$

9. $\sqrt[5]{-128x^3y^7}$

10. $(\sqrt{3} - 2\sqrt{7})(\sqrt{3} + 2\sqrt{7})$

11. $(2 - 5\sqrt{3})(2 + 6\sqrt{3})$

12. The hypotenuse of a right triangle measures 11 inches, and one leg measures 5 inches. Find the measure of the other leg.
   (a) Give its length in simplified radical form
   (b) Round the answer to the nearest thousandth.

13. (a) Write $\sqrt{-169}$ as a multiple of i
    (b) Simplify i^{87}.

Rationalize each denominator in exercises 14-16.

14. $\dfrac{3\sqrt{2}}{\sqrt{10}}$

15. $\sqrt{\dfrac{3}{5x}}$

16. $\dfrac{-2}{5 - \sqrt{7}}$

Perform the indicated operations in exercises 17-20. Express the answers in the form a + bi.

17. $(9 + i) + (3 - 4i) - (2 - 6i)$

18. $(7 - 3i)(4 - 6i)$

19. $\dfrac{2 + 6i}{2 - 3i}$

20. $\dfrac{-i}{5 - i}$

ANSWERS TO SELECTED EXERCISES

1.1   Basic Terms

1. False   **3.** True   **5.** $\{5, 6, 7, 8, \ldots\}$

7. $\{3, 2, 1, 0, -1, \ldots\}$ **9.** $\{1, -1, -3, -5, \ldots\}$ **11.** Yes

13. Possible answers include: $\{m \mid m \text{ is a multiple of 3 greater than } 0\}$.

15. Possible answers include: $\{p \mid p \text{ is an odd natural number less than or equal to } 7\}$.

17. a) $\left\{5, 7, 17, \frac{60}{2}\right\}$ b) $\left\{0, 5, 7, 17, \frac{60}{2}\right\}$ c) $\left\{-8, 0, 5, 7, 17, \frac{60}{2}\right\}$

d) $\left\{-8, -0.6, 0, \frac{3}{4}, 5, 7, \frac{17}{2}, 17, \frac{60}{2}\right\}$ e) $\left\{-\sqrt{5}, \sqrt{5}\right\}$

f) $\left\{-8, -\sqrt{5}, -0.6, 0, \frac{3}{4}, \sqrt{5}, 5, 7, \frac{17}{2}, 17, \frac{60}{2}\right\}$ g) $\left\{\frac{1}{0}\right\}$

19.

21.

23. 7   **25.** -7 **27.** -1 **29.** -4.9 **31.** 7

33. 3   **35.** True   **37.** False   **39.** $2 > -4$ **41.** $6 < 12$

43. $5 > x$ **45.** $5t - 7 \le 10$ **47.** $4 \ge 4$

49. Montana, Illinois, Maine, Oklahoma, Texas, South Carolina   **51.** Illinois

1.2   Operations on Real Numbers

1. the two numbers are additive inverses of each other   **3.** negative

5. the absolute value of the positive number is greater than the absolute value of the negative number

7. the first number is less than the second number   **9.** negative

11. 4   **13.** -18 **15.** $-\frac{3}{35}$ **17.** 0.075   **19.** 4

21. $-\frac{5}{4}$ **23.** 4.974   **25.** 11   **27.** $\frac{9}{2}$ **29.** 34

31. 168   **33.** 8.635   **35.** 3   **37.** $-\frac{20}{11}$ **39.** undefined

41. true   **43.** true   **45.** false   **47.** true   **49.** false

51. 12   **53.** 0.015625   **55.** $\frac{9}{100}$ **57.** -10

Your answers for exercises 61 and 63 may have more or less decimal places depending on the model of calculator you are using.

61. ≈ 110.199818512 **63.** ≈ 8.72811548961

65. (a) not a real number   (b) negative   **67.** 43   **69.** -119

71. 84   **73.** -8 **75.** undefined   **77.** -4

79. Nevada; increase   **81.** West Virginia; increase

| **1.3** | **Properties of Real Numbers** |

1. (b) 0   **3.** (a) −4

5. the associative properties are used to *regroup* the terms of an expression, while the commutative properties are used to change the *order* of the terms in an expression.

7. $11x$ **9.** $-6r$ **11.** cannot be simplified

13. $8p$ **15.** $3p + 3q$ **17.** $-13x + 13y$

19. $-18d - 6f$ **21.** $13x + 19$ **23.** $-2y + 8$ **25.** $-3k + 32$

27. $3p - 1$ **29.** $8p - 1$ **31.** $4z - 54$ **33.** $12x$

35. $(7 \cdot 8)r = 56r$ **37.** $14y + 7x$ **39.** $1 \cdot 8 = 8$ **41.** 0

43. $9(-3) + 9x = -27 + 9x$ **45.** 0

57. Associative property of addition   **59.** Commutative property of addition

61. Distributive property

| **1.4** | **Variables, Expressions, and Equations** |

1. (a) 13   (b) 27   **3.** (a) 22   (b) −21   **5.** (a) 4   (b) 0

7. (a) $\dfrac{5}{6}$ (b) $\dfrac{1}{2}$ **9.** (a) $-\dfrac{3}{8}$ (b) $\dfrac{9}{4}$ **11.** (a) 16   (b) 24

13. (a) $\dfrac{9}{11}$ (b) 11   **15.** (a) 2.531 (b) 4.219   **17.** $5x$

19. $x + 6$ **21.** $x - 3$ **23.** $6 - x$ **25.** $x - 5$

27. $\dfrac{14}{x}$ **29.** $8(x - 5)$ **31.** No   .**33.** $x + 6 = 16; 10$

35. $20 - \dfrac{3}{4}x = 14; 8$ **37.** $2x + 1 = 7; 3$ **39.** $3x = 2x + 10; 10$

41. expression   **43.** equation   **45.** equation   **47.** $-7 + 11 + 8; 12$

49. $(-21 + (-5)) + 12; -14$ **51.** $(-8 + (-12)) + 12; -8$

53. $(9 + (-23)) + 6; -8$ **55.** $5 - (-12); 17$ **57.** $-4 - 9; -13$

59. $(12 + (-6)) - 9; -3$ **61.** $(9 - (-3)) - 18; -6$ **63.** $7 + (-8)(3); -17$

65. $-5 - 2(-1 \cdot 8); 11$ **67.** $(1.6)(-3.7) - 8; -13.92$ **69.** $11 \cdot (6 - (-9)); 165$

71. $\dfrac{-14}{-5 + 2}; \dfrac{14}{3}$ **73.** $\dfrac{16 + (-4)}{6(-2)}; -1$ **75.** $\dfrac{-\dfrac{1}{2} \cdot \dfrac{3}{8}}{-\dfrac{4}{3}}; \dfrac{9}{64}$

77. $7x = -42; -6$ **79.** $\dfrac{x}{2} = -5; -10$

81. $x - 3 = 5; 8$ **83.** $x + 6 = -6; -12$

CH 1   **Review Exercises**

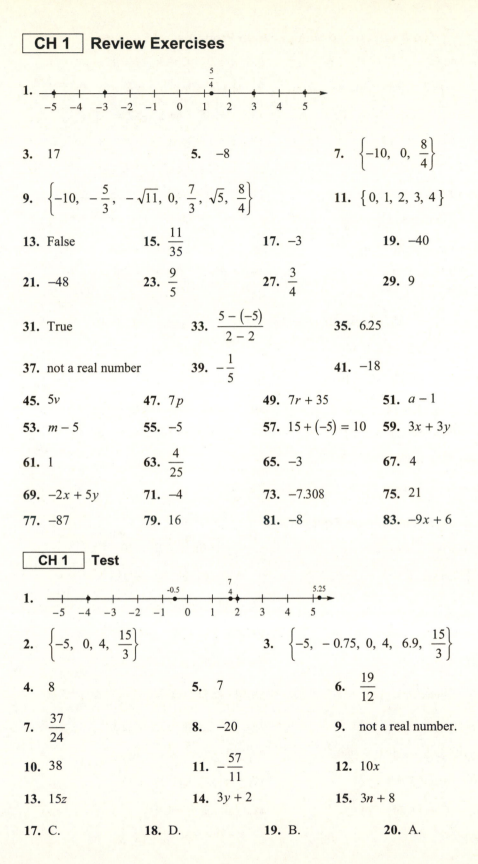

1.

3. 17   5. −8   7. $\left\{-10,\ 0,\ \dfrac{8}{4}\right\}$

9. $\left\{-10,\ -\dfrac{5}{3},\ -\sqrt{11},\ 0,\ \dfrac{7}{3},\ \sqrt{5},\ \dfrac{8}{4}\right\}$ 11. $\{\,0,\ 1,\ 2,\ 3,\ 4\,\}$

13. False   15. $\dfrac{11}{35}$ 17. −3   19. −40

21. −48   23. $\dfrac{9}{5}$ 27. $\dfrac{3}{4}$ 29. 9

31. True   33. $\dfrac{5-(-5)}{2-2}$ 35. 6.25

37. not a real number   39. $-\dfrac{1}{5}$ 41. −18

45. $5v$ 47. $7p$ 49. $7r+35$ 51. $a-1$

53. $m-5$ 55. −5   57. $15+(-5)=10$ 59. $3x+3y$

61. 1   63. $\dfrac{4}{25}$ 65. −3   67. 4

69. $-2x+5y$ 71. −4   73. −7.308   75. 21

77. −87   79. 16   81. −8   83. $-9x+6$

CH 1   **Test**

1.

2. $\left\{-5,\ 0,\ 4,\ \dfrac{15}{3}\right\}$ 3. $\left\{-5,\ -0.75,\ 0,\ 4,\ 6.9,\ \dfrac{15}{3}\right\}$

4. 8   5. 7   6. $\dfrac{19}{12}$

7. $\dfrac{37}{24}$ 8. −20   9. not a real number.

10. 38   11. $-\dfrac{57}{11}$ 12. $10x$

13. $15z$ 14. $3y+2$ 15. $3n+8$

17. C.   18. D.   19. B.   20. A.

2.1 The Addition and Multiplication Properties of Equality

1. (a) and (c)

5. $x = 22$; Check: $17 = 17$

7. $y = -19.3$; Check: $-6.2 = -6.2$

9. $z = 1$; Check: $\frac{6}{7} = \frac{6}{7}$

11. $x = 3$; Check: $26.4 = 26.4$

13. $b = -5$; Check: $-59 = -59$

15. $r = 0$; Check: $15 = 15$

17. $x = 1$; Check: $-\frac{3}{4} = -\frac{3}{4}$

19. $x = -2$; Check: $0 = 0$

23. $x = 24$; Check: $17 = 17$

25. $x = 13$; Check: $-53 = -53$

27. $x = 0$; Check: $3.5 = 3.5$

29. $s = \frac{41}{30}$; Check: $\frac{133}{180} = \frac{133}{180}$

31. $r = 1$; Check: $-9 = -9$

33. $k = 16$; Check: $-7 = -7$

35. $w = -16$; Check: $0 = 0$

37. $r = 3$; Check: $0 = 0$

39. $z = -22$; Check: $-120 = -120$

41. $x = 7$; Check: $63 = 63$

43. $k = -16$; Check: $-80 = -80$

45. $s = -\frac{19}{2} = -9.5$; Check: $-38 = -38$

47. $x = 7$; Check: $-49 = -49$

49. $x = 0$; Check: $0 = 0$

51. $p = -14$; Check: $-2 = -2$

53. $t = -19$; Check: $19 = 19$

55. $m = \frac{8}{7}$; Check: $-\frac{8}{7} = -\frac{8}{7}$

57. $x = 20$; $14 = 14$

59. $x = 12$; Check: $132 = 132$

61. $r = 7$; Check: $105 = 105$

63. $k = -72$; Check: $-8 = -8$

65. $y = \frac{45}{4}$; Check: $-5 = -5$

67. $d = -\frac{5}{21}$; Check: $\frac{5}{6} = \frac{5}{6}$

69. $a = 2.9$; Check: $-23.49 = -23.49$

71. 6

2.2 Solving Linear Equations

3. $\{2\}$; Check: $20 = 20$

5. $\{-5\}$; Check: $20 = 20$

7. $\{1\}$; Check: $8 = 8$

9. $\{21\}$; Check: $255 = 255$

11. $\varnothing$

13. {all real numbers}.

15. Solution $\{1\}$; Check: $4 = 4$

17. $\varnothing$

19. (d)

21. $\{5\}$; Check: $\frac{5}{2} = \frac{5}{2}$

23. $\{0\}$; Check: $4 = 4$

25. $\left\{-\frac{3}{5}\right\}$; Check: $\frac{7}{60} = \frac{7}{60}$

27. $\{5\}$; Check: $3.25 = 3.25$

29. $\{2\}$; Check: $12.5 = 12.5$

31. $\{4,000\}$; Check: $980 = 980$

33. $\{1\}$; Check: $40 = 40$

35. $\{0\}$; Check: $5 = 5$

37. $\{15\}$; Check: $22 = 22$

39. $\{20\}$; Check: $24 = 24$

41. {all real numbers}

43. $\varnothing$

45. $17 - q$

47. $a + 14$; $a - 7$

| **2.3** | **Problem Solving** |

1. (c) $3\frac{1}{2}$ 5. 7   7. 0

9. The height of the American Falls is 56 meters and Horseshoe Falls is 58 meters.

11. There are 4069 women and 6582 men competing.

13. One length should be 23 inches and the other should be 13 inches

15. There are 43 Democrats and 77 Republicans.

17. Mars has 2 satellites and Neptune has 8.

19. The measure of angle B and C are each 21°. The measure of angle A is 138°.

21. There are 6 ounces of cashews and 36 ounces of peanuts.   23. $p - q$

25. Yes to both parts. The angle that is equal to its supplement is 90° and the angle that is equal to its complement is 45°.

27. The angle is 75°. The supplement of 75° is 105° and the complement is 15°.

29. The angle is 37°. The supplement of 37° is 143° and the complement is 53°.

31. The angle is 75°. The supplement of 75° is 105° and the complement is 15°.

33. 63 and 64   35. 12 and 14   37. 111 and 112

39. 17 and 18   41. 9, 11, and 13

| **2.4** | **Formulas and Applications from Geometry** |

3. $P = 20$ 5. $P = 32$ 7. $A = 30$

9. $r = 30$ 11. $I = 3000$ 13. $A = 45$

15. $r \approx 2.5$ 17. $A \approx 200.96$ 19. four

21. The angles are each 102°   23. The angles are 68° and 112°.

25. $r = \dfrac{d}{t}$ 27. $p = \dfrac{I}{rt}$ 29. $b = P - a - c$

31. $m = \dfrac{y - b}{x}$ 33. $b = \dfrac{2A}{h}$ 35. $r = \dfrac{A - p}{pt}$ 37. $h = \dfrac{V}{\pi r^2}$

| **CH 2** | **Review Exercises** |

1. $\{11\}$ 3. $\{7\}$ 5. $\{17\}$ 7. $\{9\}$

9. $\{3\}$ 11. $\left\{\frac{16}{7}\right\}$ 13. $\{-12\}$ 15. $\{$all real numbers$\}$

17. $\varnothing$ 19. 24 teams participated in the 1994 World Cup.

21. 54,153 square miles is the area of Florida. 57,919 square miles is the area of Georgia.

23. The measure of the angle is 72°. The supplement of 72° is 108° and the complement is 18°.

25. $A = 70$ 27. $V \approx 33.49$

29. $h = \dfrac{2V}{b + B}$ 31. Each angle is 109°.

CH 2 | **Test**

1. $\{-7\}$ **2.** $\{24\}$ **3.** $\varnothing$ **4.** $\{10\}$

5. {all real numbers}

6. 608,827 is the resident population of Vermont. 626,932 is the resident population of Alaska and 493,782 is the resident population of Wyoming.

7. (a) $W = \dfrac{P - 2L}{2}$ or $W = \dfrac{1}{2}P - L$ (b) 13

8. Each angle is $77°$.   **9.** The measure of the angle is $64°$.

10. Solution: $\{-1\}$; Check: $-7 = -7$ **11.** Solution : $\{-1\}$; Check: $-8 = -8$

12. Solution: $\{6\}$; Check: $-27 = -27$ **13.** Solution: $\{-39\}$; Check: $-16 = -16$

14. $y = -\dfrac{2}{3}x + 4$ **15.** $n = \dfrac{A - P}{Pi}$

16. 15 calls made by Scott Hochwald's wife.

17. There are 40 pounds of gravel.   **18.** $\dfrac{r}{s}$

19. $x - 1$ **20.** 11, 13, and 15

3.1 | **Integer Exponents and Scientific Notation**

1. $(ab)^3 = a^3 b^3$ **3.** $x^2 x^5 = x^{2+5} = x^7$ **7.** x^{14}

9. x^9 **11.** $-24x^5 y^{13}$ **13.** q^7 **15.** z^4 **17.** r^{12}

19. $-27k^4$ **21.** $-140x^5$ **23.** $\dfrac{1}{3x^2 y^2}$ **25.** y^4 **29.** $\dfrac{9}{16}$

31. $\dfrac{1}{125}$ **33.** $-\dfrac{1}{25}$ **35.** $-\dfrac{1}{64}$ **37.** 7   **39.** $-\dfrac{64}{5}$

41. $\dfrac{8}{125}$ **43.** $\dfrac{1}{6}$ **45.** $-\dfrac{1}{4}$ **47.** $\dfrac{1}{27}$ **49.** $-\dfrac{1}{49}$

51. $\dfrac{1}{5184}$ **53.** 1728   **55.** $\dfrac{1}{a^8}$ **57.** $-4x^{14}$ **59.** $\dfrac{25}{y^6}$

61. $\dfrac{a^{10}}{b^{12}}$ **63.** $\dfrac{z^7}{8}$ **65.** $4p^{11}$ **67.** $-\dfrac{7}{2}y^{18}$

69. $\dfrac{b^6}{128a^7}$ **71.** $\dfrac{432}{y^{21}}$ **73.** $\dfrac{a^{13}}{18}$ **75.** $\dfrac{x^{24}}{y^{28}}$

77. 5.3×10^3 **79.** 3.45×10^{-4} **81.** 1.27×10^{-1} **83.** -4.2×10^{-4}

85. $3.21 \times 10^2 = 321$ **87.** 0.0000246   **89.** $-30,000$ **91.** 0.00000041

93. $5 \times 10^{-6} = 0.000005$ **95.** $2 \times 10^{-17} = 0.00000000000000002$

97. $1 \times 10^6 = 1,000,000$ **99.** $6 \times 10^0 = 6$

101. 63,360 inches in a mile   **103.** 32,069 hours

| **3.2** | **Addition and Subtraction of Polynomials** |

1. $4x^3 - 2x^2 + 3x + 12$ **3.** $-8p^3 - 3p^2 + 19p$

5. $-y^5 - 2y^2 + 3y + 10$ **7.** zeroth degree monomial

9. first degree binomial   **11.** third degree trinomial

13. seventh degree polynomial (none of these)

15. (d) $6x^5 - 5x^4 + 3$ **17.** $4;1$ **19.** $-12;3$

21. $1;5$ **23.** $-1;3$ **25.** $11p^4$ **27.** $12r^3$

29. $6y$ **31.** $9y^3 - y$ **33.** $11x^2 + 3x - 2$

35. $-6k^4 - 2k^3 + k^2$ **37.** $2p + 8$ **39.** $y - 12$

41. $-9p^3 + 13p^2 + 8p - 8$ **43.** $5x^5 - 2x^4 - 4x^3 + 4$

45. $2a + 7$ **49.** $25y - 3$ **51.** $-16p^2 + 11p + 3$

53. $16x^2 - 15x - 4$ **55.** $14q^2 - 5q + 12$ **57.** $2x^2 - 3x + 4$

59. $-4m^2 + 7n^2 - 16n$ **61.** $3y^4 - 14$ **63.** $9z^2 - 16z$

| **3.3** | **Multiplication of Polynomials** |

1. B.   **3.** C.   **5.** $-15x^{10}$ **7.** $-20x^{11}y^3$

9. $-8x^3 - 10x^2$ **11.** $6x^2 - 31x + 18$ **13.** $15t^2 - 58st + 11s^2$

15. $6x^2 - 5x - 4$ **17.** $9x^2 - 64$ **19.** $m^3 + 5m^2 - 36m$

21. $18z^3 - 3z^2 - 45z$ **23.** $4x^5 - 25x^3$ **25.** $2p^3 + 3p^2 + 5p + 25$

27. $12x^3 - 29x^2 + 34x - 35$ **29.** $20x^4 - 11x^3 + 2x^2 + 11x + 2$

31. $-2x^5 - 10x^4 + x^3 + 25x^2 + 10x$ **33.** $20x^4 + 34x^3 - 13x^2 - 8x - 1$

35. The product of binomials can be multiplied by the FOIL method

37. $25p^2 - 4$ **39.** $64x^2 - 1$ **41.** $49x^2 - 4y^2$

43. $9x^2 - \dfrac{4}{9}$ **45.** $9m^2 - n^4$ **47.** $49y^{10} - 9$

49. $y^2 - 18y + 81$ **51.** $9x^2 + 42x + 49$ **53.** $9m^2 - 12mn + 4n^2$

55. $a^2 - \dfrac{4}{3}ab + \dfrac{4}{9}b^2$ **59.** $4x^2 + 12xy + 9y^2 + 4x + 6y + 1$

61. $16a^2 + 8ab + b^2 - 40a - 10b + 25$ **63.** $9a^2 + 6ab + b^2 - 4$

65. $8a^3 - 2a^2b + 7ab^2 + 5b^3$ **67.** $5x^4 - 26x^3z + 20x^2z^2 + 2xz^3 - z^4$

69. $m^4 - 9m^2n^2 + 6mn^3 - n^4$ **71.** $a^4b + 2a^3b^2 - a^2b^3 - 2ab^4$

73. $y^3 + 3y^2 + 3y + 1$ **75.** $z^4 - 8z^3 + 24z^2 - 32z + 16$

3.4 | Introduction to Factoring; Special Factoring

3. $3x^3(x+1)^7$

5. $5(c+5)$

7. $7j(j^2+3)$

9. $st(1-9t)\backslash$

11. $-3p^2q^2(2q^3+p)$

13. $5x^2(3x^6-4x^3+5)$

15. $6x^2y^3(1-6x^6y+7x)$

17. $-7x^2y(2-xy^6+3xy)$

19. $3(b-2)(b+4)$

21. $(3x-2)(12-7x)$

23. $-y^7(a+b+cy+dy)$

25. $(3-x)(-4x^2+21x-22)$

27. $(2-x)(-x^2-x+4)$

29. $(2z+1)^2(8z^2+6z+5)$

31. $3x^3(-x^2+2x+3)$ and $-3x^3(x^2-2x-3)$

33. $12y^2z^3(-3y^2z^2-z+2y^3)$ and $-12y^2z^3(3y^2z^2+z-2y^3)$

35. $(a+4b)(b+d)$ **37.** $(1+2m)(3k+7p)$ **39.** $(x-3)(x+4)$

41. $(a+c)(a-3b)$ **43.** $(x+4)(3x-5)$ **45.** $(4x+5y)(y-5x)$

47. $(2-b)(3-a)$ **49.** $(7x+2)(x^3-2)$ **51.** $(1-x)(1-4y)$

53. (a) B.   (b) D.   (c) A.   (d) C.

55. sum of cubes: (d); difference of cubes: (a) and (b); neither: (c)

57. $(x-4)(x+4)$

59. $(6x-7)(6x+7)$

61. $(5x-6y)(5x+6y)$

63. $3(x-3y)(x+3y)(x^2+9y^2)$

65. $(x+y-4)(x+y-4)$

67. $(7-x-2y)(7+x+2y)$

69. $4ab$

71. $(k-4)^2$

73. $(3a-b)^2$

75. $(7x-1-y)(7x-1+y)$

77. $(4x-3-y)(4x-3+y)$

79. $6(x+3)^2$

81. $(a+b-6)^2$

83. $(4a+b)(16a^2-4ab+b^2)$

85. $(10x-3y)(100x^2+30xy+9y^2)$ **87.** $(9a^3-1)(9a^3+1)$

89. $(x+y-1)(x^2+2xy+y^2+x+y+1)$

91. $(9a^2-8b^5)(81a^4+72a^2b^5+64b^{10})$

3.5   Factoring Trinomials

1. (a) D. (b) F. (c) A. (d) B (e) C. (f) E.   **3.** $(y + 12)(y + 2)$

5. $(r - 6)(r - 5)$ **7.** $(y + 6)(y - 4)$ **9.** $(ab + 5)(ab + 1)$

11. $(r - 8s)(r - 2s)$ **13.** $(s - 2t)(s + 9t)$ **15.** $(3y + 4)(5y + 9)$

17. $(6x + 5)(2x - 9)$ **19.** $(3j + 8)(5j - 6)$ **21.** $(3a - 4b)(5a - 2b)$

23. $(7r + 3s)^2$ **25.** $(7c - d)^2$ **27.** $(3g + 5h)(2g - h)$

29. $2(p - 11)(p - 2)$ **31.** $6(a - 11)(a - 2)$ **33.** $x(x - 4)(x + 3)$

35. $3x^2(x + 3)(x + 2)$ **37.** $2xy(3x + 8)(2x + 1)$ **39.** $2y^2(10x + 3)^2$

41. $4y(6x + 5y)(2x - 9y)$ **43.** $2z(5a - 2c)(3a - 4c)$ **45.** $10p(3p + 2)^2$

47. $-(x + 9)(x - 6)$ **49.** $-(5k - 6)(2k + 3)$ **51.** $-2x(7x - 1)(5x + 3)$

3.6   Division of Polynomials

1. quotient; exponents   **3.** descending powers of the variables

5. $5y + 3 - \dfrac{4}{y}$ **7.** $\dfrac{3}{2}x - \dfrac{5}{2} + \dfrac{1}{x}$

9. $\dfrac{3x^2}{2} - \dfrac{6x}{7y} + \dfrac{4}{7}$ **11.** $2xz + \dfrac{2y}{5z} + \dfrac{1}{xy}$

13. $x^2 - 3x + 4$ **15.** $y - 4$

17. $2m + 3$ **19.** $2x^2 - 1$

21. $t^2 + 6t - 8$ **23.** $4y^2 - 2y + 5 - \dfrac{8}{2y + 3}$

25. $3b^2 - 5b - 2$ **27.** $2p^2 + 3p + 1 - \dfrac{2}{p^2 - 2p + 3}$

29. $p^2 + 2p + 4$

CH 3   Review Exercises

1. $-\dfrac{8y^{10}}{x}$ **3.** $\dfrac{4h^7}{k^3}$ **5.** 16   **7.** -8

9. $\frac{16}{9}$ **11.** $\frac{3}{4}$ **13.** 0   **15.** $\frac{1}{64}$

17. $\dfrac{y^6}{x^3}$ **19.** $36p^{22}$ **21.** $\dfrac{1}{36k^6}$ **23.** Yes

27. 1.75×10^{-7} **29.** 250,000   **31.** 186,000   **33.** 1600

35. 3,000,000   **37.** 17   **39.** 14

41. (a)$-9a^7 + 2a^5$ (b)binomial (c)degree 7   **43.** (a)l^9 (b)monomial (c)degree 9

45 $-x^2 - x - 10$ **47.** $-2r^3 - 9r^2 - r + 12$

49. $10x^2 + 7x + 23$ **51.** $6x^2 - 19x - 7$

53. $14s^2 + 23st - 30t^2$ **55.** $8x^3 - 12x - 5$

57. $4k^2 - 25$ **59.** $9c^2 + 6c + 1$

61. $2x^2 - 4x = 2x(x - 2)$ **63.** $3ab(2a - 3a^2b + 4b)$

65. $3(a + 3)(a - 2)$ **67.** $(a - b)(5m - n)$

69. $(n + 8)(p + r)$ **71.** $(3x + 4)(2x + 5)$ **73.** $(7y - 1)(2y + 7)$

75. $9x^2 - 6xy - 8y^2 = (3x - 4y)(3x + 2y)$

77. $2x(7x - 2)(3x - 5)$ **81.** $(5x - 4)(5x + 4)$

83. $(7y + 5)^2$ **85.** $(x + 5)(x^2 - 5x + 25)$

87. $(x^2 + 2)(x^4 - 2x^2 + 4)$ **89.** $2t(3s^2 + t^2)$

91. $2x^3 - \dfrac{3x^2}{2} - \dfrac{1}{x}$ **93.** $6x^3 - 19x^2 + 5$

CH 3   **Test**

1.   (a) E.   (b) C.   (c) B.   (d) C.   (e) D.   (f) F.   (g) A.   (h) G.   (i) E.

2. $\dfrac{2x^3}{9y^3}$ **3.** $\dfrac{4h^{10}}{k}$ **4.** $\dfrac{16}{27j^{42}k^8}$

5. $-\dfrac{z^{12}}{8x^6}$ **6.**   (a) 1,013,400   (b) 1,500,000,000

7. $3x^3 + 8x^2 - 5x$ **8.** $15x^2 + 8x - 63$

9. $10x^3 - 13x^2 - 2x + 3$ **10.** $9m^2 - 16e^2$

11. $25a^2 - 90a + 81$ **12.** $11a^3(11 - 12a)$

13. $(3p - 2)(p - 5)$ **14.** $(3x + 11y)(2x - 3y)$

15. $(3x - 4)^2$ **16.** $(3j - 7k)(3j + 7k)$

17. $(x + 6)(x^2 - 6x + 36)$ **18.** $4xy(7x - 2)(3x - 5)$

19. $\dfrac{3x^2}{2} - 3xy - \dfrac{9}{2y^2}$ **20.** $3x^2 - 6x - 2 + \dfrac{3}{3x - 1}$

| **4.1** | **Fractions** |

1. True 3. False 5. $\frac{1}{2}$ 7. $\frac{3}{8}$ 9. $\frac{3}{8}$ 11. $\frac{13}{15}$ 13. (c)

15. $\frac{8}{15}$ 17. $\frac{3}{14}$ 19. $\frac{4}{9}$ 21. $\frac{27}{8}$ 23. $\frac{2}{7}$ 25. $\frac{5}{8}$ 27. $\frac{4}{3}$

29. $\frac{3}{16}$ 33. $\frac{1}{2}$ 35. $\frac{13}{16}$ 37. $\frac{7}{8}$ 39. $\frac{5}{6}$ 41. $\frac{44}{45}$ 43. $\frac{35}{24}$

| **4.2** | **Rational Expressions: Multiplication and Division** |

1. B.   3. C.   5. D.   7. $\frac{4y^5}{3x^6}$ 9. $\frac{x+6}{x+9}$ 11. $\frac{x(x+2)}{2(x-2)}$

13. cannot be reduced further 15. $\frac{3}{4}$ 17. $\frac{t}{6}$

19. $\frac{6}{y-3}$ 21. $\frac{c-5}{c-7}$ 23. $\frac{2k+9}{2k-3}$

25. $\frac{1}{g^2+2g+4}$ 27. $\frac{2x+3y}{2(3x-y)}$ 29. $\frac{x+y}{x-y}$

31. (b)   29. -1 31. $-(a+b)$

33. $-\frac{3a-1}{3a+1}$ 35. $-\frac{3}{2}$ 37. cannot be reduced further.

41. $\frac{xy^2}{4}$ 43. $\frac{25a^{14}}{b^{11}}$ 45. $\frac{3y^{11}}{2}$

47. $\frac{9x}{20}$ 49. $-\frac{2(x+4)}{x}$ 51. $-\frac{d(d+8)}{d+5}$

53. $-\frac{85}{6}$ 55. $\frac{(x+6)(x+7)}{x-9}$ 57. $\frac{(3x+4y)(2x+y)}{(x-3y)(3x-y)}$

59. $\frac{(s+2t)(s-2t)}{(s-8t)(2s-9t)}$ 61. $\frac{1}{8x-1}$ 63. $\frac{(x+3)(7x+1)}{(x+6)}$

| **4.3** | **Addition and Subtraction of Rational Expressions** |

1. $\frac{9}{20}$ 3. No   5. -1 7. $\frac{5}{c}$ 9. $\frac{4}{x^2}$

11. 1   13. $d-4$ 15. $\frac{5}{x+1}$ 17. $a-2$ 19. $48x^2y^9z^6$

21. $c(c-7)$ 23. $4(x+4)$ 25. $24(p-2)$ 27. $(a-b)(a+b)$

29. $y(y+1)(y-7)$ 31. $p(p+3)(p-6)$ 33. $(s+2)(3s-5)(s-5)$

35. $2x(x-2)(x+2)$ 37. $12g^5(g-2)$ 39. Yes

41. $\frac{11}{4t}$ 43. $\frac{20y-21x}{24x^2y^2}$ 45. $\frac{16}{t(t-4)}$

47. $\dfrac{a(5a-12)}{(a-2)(a-3)}$ **49.** $\dfrac{d-4}{d-6}$ **51.** $\dfrac{2(2a+3b)}{a-b}$ **53.** $\dfrac{x+8}{x+3}$

55. $\dfrac{7}{f}$ **57.** $\dfrac{2x+1}{x+2}$ **59.** $\dfrac{2(v-5)}{v^2-2v+4}$ **61.** $\dfrac{x-10}{x}$

63. $\dfrac{2(2c^2-15c+30)}{(c-3)^2}$ **65.** $\dfrac{x^2+6xy-2y^2+2x+3y}{(x-y)(x-2y)(x+5y)}$

67. $\dfrac{-3a^2-6ab-2b^2}{(a-b)(2a-3b)(a+b)}$

4.4   Complex Fractions

3. $\dfrac{3x}{2(x-5)}$ **5.** $\dfrac{2(3z-8)}{2z+1}$ **7.** $\dfrac{5x}{8y^2}$ **9.** $\dfrac{3-d}{2d+5}$

11. $\dfrac{y+x}{y-x}$ **13.** $\dfrac{2x}{21}$ **15.** $a+3b$ **17.** $\dfrac{5d}{2}$

19. $\dfrac{z^2-7z-12}{z^2+3z-9}$ **21.** $\dfrac{x^4y^4}{y^4+x^4}$ **23.** $\dfrac{y+x}{xy}$ **25.** $\dfrac{xy}{y-x}$

CH 4   Review Exercises

1. a) $\dfrac{2}{3}$ b) $\dfrac{3}{4}$ c) $\dfrac{7}{17}$ **3.** a) $\dfrac{1}{4}$ b) $\dfrac{43}{48}$ c) 1   **5.** $\dfrac{8x}{7y^4}$

7. $\dfrac{2a+9b}{2a-3b}$ **11.** $-\dfrac{5(g+5)}{g}$ **13.** 1   **17.** $16f^2(2f+1)$

19. $\dfrac{9-4c}{4c^2}$ **21.** $\dfrac{13}{10(2a+3)}$ **23.** $\dfrac{5+2b}{2-5b}$ **25.** $\dfrac{1}{5y+7x}$

CH 4   Test

1. a) $\dfrac{9}{17}$ b) $\dfrac{7}{13}$ **2.** a) $\dfrac{15}{143}$ b) 15   **3.** a) $\dfrac{34}{63}$ b) $\dfrac{8}{15}$ **4.** $3xy$

5. $\dfrac{3x-1}{2x-1}$ **6.** $\dfrac{2a+9b}{2(2a+b)}$ **7.** $-\dfrac{5(1+3g)}{3g+7}$ **8.** $\dfrac{3}{4}$

9. $\dfrac{2(f-1)}{3(f+9)}$ **10.** $-\dfrac{5z+1}{5z-2}$ or $\dfrac{5z+1}{2-5z}$

11. -1 **12.** $12x^3y^7$

13. $24r(r-1)(r+1)(2r+1)$ **14.** $9(3x+5)(2x+1)^2$

15. $\dfrac{2(3z+2)}{4z+7}$ **16.** $\dfrac{11n^2+20}{2n(n-2)(n+4)}$ **17.** $\dfrac{17x+8}{(3x+1)(x-2)^2}$

18. $-\dfrac{39}{p+8}$ **19.** $\dfrac{3}{8}$ **20.** $\dfrac{b}{2a+3b}$

5.1   Evaluating Roots

1. False. Zero is nonnegative and has only one square root, namely zero.

3. True

5. True

7. 2 and -2

9. 14 and -14

11. $\frac{2}{5}$ and $-\frac{2}{5}$

13. 50 and -50

15. 6

17. -19

19. $-\frac{2}{3}$

21. not a real number

23. 11.091

25. 2.675

27. 11.253

29. 30

31. 34

33. 7.810

35. 2.080

37. 4.580

39. $c = 40$

41. $b = 6$

43. $c = \sqrt{130} \approx 11.402$

45. $40\,\text{cm}$

47. 100 feet

5.2   Multiplication and Division of Radicals

1. 4

3. $6\sqrt{2}$

5. 21

7. $\sqrt{6x}$

9. $3\sqrt{7}$

11. $6\sqrt{3}$

13. $10\sqrt{2}$

15. $-20\sqrt{2}$

17. $26\sqrt{3}$

19. $2\sqrt{7}$

21. 12

23. $7\sqrt{6}$

27. $\dfrac{5}{14}$

29. 2

31. 6

33. z

35. f^3

37. $20x^5$

39. $2\sqrt[3]{7}$

41. $3\sqrt[3]{5}$

43. $2\sqrt[4]{6}$

45. $\dfrac{3}{4}$

47. radical expressions

49. fraction

51. $2\sqrt{5}$

53. $-\dfrac{\sqrt{65}}{5}$

55. $\dfrac{2\sqrt{30}}{3}$

57. $\dfrac{10\sqrt{3}}{9}$

59. $\dfrac{21\sqrt{2}}{2}$

61. $\sqrt{2}$

63. $\dfrac{\sqrt{2}}{2}$

65. $\dfrac{\sqrt{35}}{7}$

67. $\dfrac{\sqrt{55}}{5}$

69. $\dfrac{17\sqrt{21}}{63}$

71. $\dfrac{1}{6}$

73. 1

75. $\dfrac{\sqrt{2x}}{x}$

77. $\dfrac{2s\sqrt{st}}{t}$

79. $\dfrac{x\sqrt{3xy}}{y}$

81. $\dfrac{5ab^2\sqrt{3bc}}{3c}$

5.3   Addition and Subtraction of Radicals

1. distributive

3. radicals

5. $15\sqrt{6}$

7. $10\sqrt{13}$

9. $4\sqrt{2}$

11. $23\sqrt{2}$

13. $18\sqrt{2}$

15. $-23\sqrt{6}$

17. $6\sqrt{5}$

19. $42\sqrt{2}$

21. $104\sqrt{3}$

23. $\sqrt{6}$

27. $12\sqrt{2}$

29. $-3\sqrt{x}$

31. $6a\sqrt{3}$

33. $32x\sqrt{10}$

35. $21\sqrt{3k}$

37. $20b\sqrt{5a}$

39. $-8\sqrt[3]{2}$

41. $28s\sqrt[3]{r^2}$

43. 0

5.4   Simplifying Radical Expressions

1. 11   **3.** 9   **5.** 12   **7.** 1

9. $16\sqrt{3}$ **11.** $-18\sqrt{6}$ **13.** $\sqrt{6} - \sqrt{21}$ **15.** $3\sqrt{15} - 30$

17. $-10\sqrt{5}$ **19.** $54 + 9\sqrt{10}$ **21.** $52 + 13\sqrt{14}$ **23.** $81 + 36\sqrt{5}$

25. 1   **27.** 4   **29.** $2\sqrt{3} + 2 - 3\sqrt{2} - \sqrt{6}$

31. $18\sqrt{2} - 18$ **33.** $159 + 28\sqrt{5}$ **37.** $2\left(2 - \sqrt{3}\right)$ **39.** $\dfrac{6\left(8 + \sqrt{6}\right)}{29}$

41. $2\sqrt{3} - 3$ **43.** $\dfrac{\sqrt{15} - 3}{2}$ **45.** $10 - 4\sqrt{5}$

47. $\dfrac{4\sqrt{2} + \sqrt{10} + 12 + 3\sqrt{5}}{11}$ **49.** $2\sqrt{2} + 3$

51. $\dfrac{2\sqrt{7} - 4}{3}$ **53.** $3 - \sqrt{3}$

55. $x\sqrt{6} + 3\sqrt{x} + 2\sqrt{6x} + 6$ **57.** $y\sqrt{14} - 7\sqrt{y} + 2\sqrt{7y} - 7\sqrt{2}$

59. $3 + 4\sqrt[3]{9}$ **61.** $9\sqrt[4]{8} - 18$ **63.** $\sqrt[3]{4} - \sqrt[3]{9}$ **65.** 5

5.5   Fractional Exponents

1. (a) -9 **3.** (c)3   **5.** 6   **7.** 5   **9.** 3   **11.** 3

13. 32   **15.** 32   **17.** 8   **19.** -729 **21.** -81 **23.** $\dfrac{1}{36}$

25. $\dfrac{1}{3}$ **27.** $-\dfrac{1}{5}$ **29.** 9   **31.** $\dfrac{1}{3}$ **33.** $\dfrac{1}{10^{1/2}}$ **35.** $5^{1/5}$

37. 36   **39.** $14^{1/2}$ **41.** $\dfrac{216}{343}$ **43.** $\dfrac{1}{3^{3/5}}$ **45.** $\dfrac{1}{7^{1/3}}$ **47.** $x^{8/5}$

49. $a^{10}b^{4/7}$ **51.** $\dfrac{x^{4/7}}{y^{24/35}}$ **53.** 3   **55.** 4   **57.** $\sqrt{y}$ **59.** $\sqrt[3]{s^2}$

5.6   Complex Numbers

1. i **3.** 1   **5.** i **7.** $12i$ **9.** $-13i$

11. $i\sqrt{10}$ **13.** $2i\sqrt{5}$ **15.** -14 **17.** -21 **19.** $\sqrt{3}$

21. $2i$ **25.** $5 + 3i$ **27.** $-10 - 3i$ **29.** $3 + 9i$

31. $23 - 2i$ **33.** $12 + 0i$ or 12   **35.** $25 + 4i$ **37.** $-2 - 3i$

39. $7 + 6i$ **41.** -14 **43.** -12 **45.** $-18 - 4i$

47. $53 + 26i$ **49.** $-45 - 28i$ **51.** 4   **53.** 85

55. 39   **57.** $3 - 5i$ **59.** $2 + 2i$ **61.** $-10 + 6i$

63. $\frac{38}{29} - \frac{8}{29}i$ **65.** -1 **67.** $-i$ **69.** $-i$

CH 5   Review Exercises

1. 1 and −1   **3.** 8 and −8   **5.** 24 and −24   **7.** 5

9. 5   **11.** not a real number   **13.** $\dfrac{9}{17}$

15. x must be zero or a positive real number.   **17.** irrational; 5.099

19. rational; −28   **21.** $2\sqrt{35}$ **23.** $5\sqrt{10}$ **25.** 15

27. $15\sqrt{6}$ **29.** $-\dfrac{13}{19}$ **31.** $\dfrac{\sqrt{7}}{11}$ **33.** $\dfrac{\sqrt{35}}{4}$

35. 4   **37.** d **39.** a^6 **41.** $a^6 b^8 \sqrt{ab}$

45. $18\sqrt{3}$ **47.** $19\sqrt{3}$ **49.** $2\sqrt{2}$ **51.** $-\sqrt{10}$

53. $21\sqrt{y}$ **55.** $11g^2\sqrt{2h}$ **57.** $\sqrt{3}$ **59.** $\dfrac{\sqrt{39}}{13}$

61. $2\sqrt{3}$ **63.** $\dfrac{x\sqrt{y}}{5y}$ **65.** $\dfrac{\sqrt{3st}}{6t^2}$ **67.** $-\sqrt{15} - 10$

69. $32 - 11\sqrt{7}$ **71.** −53   **73.** $\dfrac{15 - 3\sqrt{3}}{22}$

75. $\dfrac{5\sqrt{2} - 3\sqrt{10}}{4}$ or $\dfrac{3\sqrt{10} - 5\sqrt{2}}{4}$ **77.** $-\dfrac{\sqrt{15} + 3\sqrt{3} + \sqrt{5} + 3}{4}$

79. $\dfrac{8 + 5\sqrt{2}}{2}$ **81.** $\dfrac{\sqrt{5} - \sqrt{2}}{5}$ **83.** −4   **85.** $\dfrac{1}{11}$

87. 13   **89.** $9 - i$ **91.** $-25 + 10i$ **93.** 82

CH 5   Test

1. 18 and −18   **2.** 9   **3.** $-50\sqrt{3}$ **4.** $\dfrac{5\sqrt{6}}{9}$

5. $2\sqrt[4]{2}$ **6.** $5\sqrt{3}$ **7.** $-28\sqrt{5}$ **8.** $-90x\sqrt{3}$

9. $-2y\sqrt[5]{4x^3 y^2}$ **10.** −25   **11.** $-86 + 2\sqrt{3}$

12. (a) $4\sqrt{6}$ (b) 9.798   **13.** (a) $13i$ (b) $-i$ **14.** $\dfrac{6\sqrt{5}}{10}$

15. $\dfrac{\sqrt{15x}}{5x}$ **16.** $-\dfrac{5 + \sqrt{7}}{9}$ **17.** $10 + 3i$

18. $10 - 54i$ **19.** $-\frac{14}{13} + \frac{18}{13}i$ **20.** $\frac{1}{26} - \frac{5}{26}i$